Edward Hamilton

The River-Side Naturalist

Notes on the various forms of life met with either in, on, or by the water, or in its immediate vicinity. Illus. with numerous woodcuts

Edward Hamilton

The River-Side Naturalist
Notes on the various forms of life met with either in, on, or by the water, or in its immediate vicinity. Illus. with numerous woodcuts

ISBN/EAN: 9783337095123

Printed in Europe, USA, Canada, Australia, Japan

Cover: Foto ©Lupo / pixelio.de

More available books at **www.hansebooks.com**

THE
River-Side Naturalist

NOTES ON THE VARIOUS FORMS OF LIFE MET WITH
EITHER IN, ON, OR BY THE WATER, OR IN
ITS IMMEDIATE VICINITY.

BY

EDWARD HAMILTON, M.D., F.L.S., F.Z.S.,
AUTHOR OF
"RECOLLECTIONS OF FLY-FISHING FOR SALMON, TROUT, AND GRAYLING."

Illustrated with Numerous Woodcuts.

LONDON:
SAMPSON LOW, MARSTON, SEARLE, & RIVINGTON,
LIMITED,
St. Dunstan's House,
FETTER LANE, FLEET STREET, E.C.
1890.

[*All rights reserved.*]

INTRODUCTION.

SOME writer on Natural History whose name we forget says, that in every one of us there is an innate love of Nature in her purest and simplest phases. Yet how few there are who, having this love and enjoying thoroughly the beauties which meet their gaze when wandering by the water-side, know anything of the various forms of animal life which are so profusely distributed in every direction! How many are there who know the difference between what is called the water-rat and the rat of the barn or corn-stack?—between the weasel and the stoat?—the field-mouse and the house-mouse? The flowers which carpet the meadows and marshes with blue and yellow in spring-tide, with purple and gold in autumn, are to many nothing but flowers; the insects which hum and fly around us, nothing but beetles or flies, butterflies or moths.

A little more knowledge of Nature's handicraft, so as to be able to distinguish the various species, would make the hours pass more pleasantly to any one, whatever might be his occupation by the river-side, but more especially to the angler, when either waiting for his float to bob, or sitting on a rail expecting the "rise."

Ray, in his work on the "Wisdom of God in the Creation," in writing of the wonders of the elements, of the birds, insects, and other living beings, says: "Man is commanded to consider them, and particularly to observe and take notice of their various structures, ends, and uses, and to give God the praise;" and adds, that the reason why so many kinds of creatures were made "might be to exercise the contemplative faculty of man;" and we have only to exercise the faculty of observation to find ourselves in the presence of innumerable objects of interest. We can be "far from the madding crowd," yet in the midst of living beings. Rest for a moment on that old and rotten tree, levelled to the ground by a passing storm; strip off a piece of the bark, it will be found to be teeming with life. Sit on that old boulder—relic, may be, of the glacier age—now covered with moss and lichen, and turn over the dead leaves at your feet; beetles and other insects of various forms and hues will scuttle away to seek some fresh hiding-place. Cast your eyes upwards; the air above is peopled with winged creatures, the trees and hedgerows resound with the "hum of bees, the voice of birds."

Go where you will—on the placid waters of the meres and lakes, by the rushing rivers or babbling brooks—animated nature is above and around you. Birds are singing in the air, resting in the bushes, creeping or hiding in the reeds. Listen to the warning *Churr-churr* of the sedge-warbler—hark to the carol of the lark, a speck in the blue ether—watch the rapid flight of the swift, now skimming the surface of the water in front of us, and now far away over the distant meadows. New objects, as Ray says, afford us great delight, especially if discovered by our own industry, and provide us employment most delightful and agreeable to our nature and inclinations.

In describing the various objects in the following pages, our aim has been to bring them before the reader as they may be met with, either in or on the water, or in its imme-

INTRODUCTION. vii

diate neighbourhood, and without regard to any systematic arrangement except so far as to separate the quadrupeds from the birds, the reptiles from the fishes, and so on.

A good binocular field-glass, to bring the objects in closer approximation to the eye; a pocket magnifier, to examine the insects and other small objects; and a note-book, to record results, are strongly recommended to all who take an interest in the natural history of the waterside.

These notes, now considerably enlarged and with many new illustrations, were originally published in the *Fishing Gazette* under the title of "What we See when we go a-Fishing," and were suggested by the following circumstance:—One fine September morning, on grayling-fishing intent, when about to commence operations upon a very famous shallow of a very famous river, a flight of goldfinches on their autumn migration flew over our heads and settled on a bunch of thistles on the opposite bank. On calling our friend's attention to these beautiful birds, we found that he was almost entirely ignorant of the various forms of animal life so constantly met with where water abounds; and from some further remarks made by him on this subject, we ventured into print.

CONTENTS.

CHAPTER I.
THE QUADRUPEDS.

	PAGES
The Water-Vole—The Otter—The Weasel—The Stoat—The Shrew-Mouse—The Water-Shrew—The Bats	1–20

CHAPTER II.
THE BIRDS.

The Kingfisher—The Sedge-Warbler—The Reed-Warbler—The Reed-Sparrow—The Bearded Titmouse—The Moor-Hen—The Coot—The Water-Rail—The Water-Ouzel—The Dabchick—The Mallard—The Swan—The Red-Breasted Merganser—The Teal . . . 21–55

CHAPTER III.

The Lapwing—The Snipe—The Heron—The Bittern—The Common Sandpiper—The Water-Wagtail . . . 56–69

CHAPTER IV.

The Swift—The Swallow—The Martin . 70–79

CHAPTER V.

The Lark—The Starling—The Cuckoo—The Song-Thrush—The Blackbird—The Nightingale—The Willow-Warbler—The White-Throat—The Wren—The Red-Breast—The Black-Cap—The Wheat-Ear—The Spotted Fly-Catcher—The Redstart—The Hedge-Sparrow . 80–116

CONTENTS.

CHAPTER VI.

The Goldfinch—The Chaffinch—The House-Sparrow—The Yellow-Hammer—The Wry-Neck—The Green Woodpecker—The Great Spotted Woodpecker . . . 117–128

CHAPTER VII.

The Marsh Titmouse—The Coal Titmouse—The Long-Tailed Titmouse—The Great Titmouse—The Blue Titmouse 129–135

CHAPTER VIII.

The Rook—The Jackdaw—The Magpie and Jay—The Wood-Pigeon—The Dotterel—The Curlew—The Sparrow-Hawk—The Kestrel Hawk 136–156

CHAPTER IX.

The Barn-Owl—The Brown or Tawny Owl—The Nightjar . 157–165

CHAPTER X.

THE REPTILES.

The Lizards—The Slow-Worm—The Snakes—The Frogs—The Toads—The Newts 166–177

CHAPTER XI.

THE FISHES.

Their Mental Intelligence and Powers of Movement . . 178–192

CHAPTER XII.

The Perch—The Pope or Ruffe—The Basse—The River Bull-Head—The Sticklebacks 193–203

CONTENTS.

CHAPTER XIII.

The PHYSOSTOMI: The Carp—The Barbel—The Gudgeon—The Loaches—The Tench—The Crucian Carp—The Bream—The Roach—The Chub—The Rudd—The Dace—The Bleak—The Minnow—The Shad . 204–230

CHAPTER XIV.

The *Salmonidæ*—The Salmon—The Sea-Trout—The River-Trout—The Grayling—The Char—The Smelt . . 231–279

CHAPTER XV.

The Pike 280–285

CHAPTER XVI.

The Eels—The Lampreys 286–292

CHAPTER XVII.

MOLLUSCS, CRUSTACEANS, AND ANNELIDÆ
(EARTH-WORMS).

Slugs—Snails—The Crayfish—The Fresh-Water Shrimp . 293–298

CHAPTER XVIII.

THE INSECTS.

The *Gyrinus Natator*—The Great Water-Beetle—The Water-Boatman—The Water-Bug—The Devil's Coach-Horse—The Common Dor-Beetle—The Peacock-Fly—The Fern-Fly—The Marlow Buzz—The May-Bug—The *Blethisa Multipunctata*—The *Odocantha Melanusa* 299–308

CONTENTS.

CHAPTER XIX.

NEUROPTERA: The Great Dragon-Fly—The Dragon-Fly or Horse-Singer—The *Libellula Puella* and *L. cæruleus*—The *Phryganidæ*: The Grannom—The Red or Cinnamon Sedge—The Silver Sedge—The Silver Horns—The Sand-Fly—The Stone-Flies—The Scorpion-Flies—The Alder-Fly—The Chantry—The *Perlidæ*: The Yellow Sally—The Willow-Flies—The Grasshoppers 309–323

CHAPTER XX.

Ephemeridæ: The May-Flies, &c. 324–346

CHAPTER XXI.

The HYMENOPTERA: The Ants—The Saw-Flies—The Governor—The HEMIPTERA: The Water-Cricket—The Water-Scorpion—The Frog-Hopper—The DIPTERA: The Golden Dun Midge—The Cow-Dung Fly—The Gravel-Bed or Spider Fly—The Black Gnat—The Fisherman's Curse—The Bluebottle—The Common House-Fly—The Down-Hill Fly—LEPIDOPTERA: Butterflies—Moths 347–355

CHAPTER XXII.

OF THE EARTH-WORMS (ANNELIDÆ).

The Earth-Worm—The Marsh-Worm—The Black-Head—The Brandling—*Lumbricus fœtidus*—Leeches . . 356–358

CHAPTER XXIII.

RIVERSIDE FLOWERS.

The Buttercup—The Daisy—The Pale Primrose—The Cowslip—The Violet—The Dog-Violet—The Wood-Sorrell—The Wild Hyacinth—The Common Daffodil—The Venus's Catchfly—The Bog Asphodel—The Grass of Parnassus—The Common Marsh Marigold—

The Spiked Water-Milfoil—The Common Butter-Bur—The Great Sedge—The Lesser Sedge—The Marsh Pennywort—The Water-Crowfoot—The Water Sweet-Grass—The Comfrey—The Bur-Marigold—The Nodding Bur-Marigold—The Celery-Leaved Crowfoot—The Great Wild Valerian—The Sweet-Flag—The Water-Violet—The Great Bladderwort—Pond-Weeds—The Broad-Leaved Pond-Weed—The Flowering Rush—The Mare's Tail—The Great Water-Horsetail—The Yellow Water-Iris—The Forget-Me-Not—The Common Brook Lime—The Gipsy-Wort, Water-Horehound—The Common Buck-Bean—The Common Meadow-Rue—The Creeping or Marsh Spike Rush—The Frog-Bit—The Club-Rush—The Common Hornwort—Watercress—The Bur-Reeds—The Water Figwort—The Water-Dropworts—The Common Yellow Water-Lily—The Large White Water-Lily—The Common Duckweed—The Great Water-Dock—The Purple Loosestrife—The Meadow-Sweet—The Willow Herbs—The Common Reed—The Red Canary-Grass—The Reed Meadow-Grass—The Floating Meadow-Grass—The Great Cat's Tail—The Lesser Cat's Tail—The Marshwort—The Water-Hemlock—The Broad-Leaved Water-Parsnep—The Common Arrow-Head—The Plantains—The Common Horn Pond-Weed—The Hemp or Water-Agrimony—The Common Flea-Bane—The Marsh Woundwort—The Tall Red Mint—The Hairy Mint—The Water-Thyme 359–394

LIST OF ILLUSTRATIONS.

QUADRUPEDS.

	PAGE		PAGE
The Water-Vole	1	The Stoat	13
Heads of Water-Vole and Rat	3	The Shrew-Mouse	14
The Otter	6	The Water-Shrew	16
The Weasel	10	Head of the Great Bat	20

BIRDS.

	PAGE		PAGE
The Kingfisher	23	Nest of Reed-Warbler	69
The Sedge-Warbler	26	The Swift	71
The Reed-Warbler	28	The Swallow	73
The Reed-Sparrow	29	The Martin	77
The Bearded Titmouse	30	The Skylark	81
The Moor-Hen	31	The Tree Pipit	84
The Coot	34	The Meadow Pipit	84
The Water-Rail	37	The Starling	85
The Water-Ouzel	38	The Cuckoo	87
The Little Grebe. Dabchick	44	The Song-Thrush	93
The Mallard	47	The Blackbird	95
Heads of the Mute Swan and the Wild Swan	51	The Nightingale	97
		The Willow-Warbler	100
Head of Merganser	53	The White-Throat	101
The Teal	55	The Wren	103
The Lapwing	56	The Redbreast	105
The Snipe	59	The Black-Cap	109
The Heron	61	The Wheat-Ear	111
Head of Bittern	64	The Redstart	113
The Common Sandpiper	65	The Hedge-Sparrow	115
The Water-Wagtail	67	The Whinchat	116
Heads of Pied, White, Yellow, and Grey Wagtail	68	The Goldfinch	118
		The Chaffinch	119

xvi LIST OF ILLUSTRATIONS.

	PAGE		PAGE
Head of the Yellow-Hammer	122	The Rook	137
The Wryneck	123	Heads of Rook and Carrion Crow	146
The Green Woodpecker	125	The Jackdaw	147
Diagram to show the Tongue of the Woodpecker	126	Head of the Wood-Pigeon	149
		The Dotterel	150
The Great Spotted Woodpecker	128	The Curlew	152
		Head of the Sparrow-Hawk	153
The Marsh Titmouse	129	The Kestrel Hawk	155
The Coal Titmouse	131	Head of the Barn-Owl	159
The Long-Tailed Titmouse	132	Head of the Tawny Owl	160
The Ox-Eye or Great Titmouse	133	The Nightjar	162
		Head and Foot of the Nightjar	164
The Blue Titmouse	134		

REPTILES.

The Lizards	167	The Frog	172
Head of the Slow-Worm	169	The Toad	174
Common Viper or Adder	170	The Newts	176
Ringed or Common Snake	170		

FISHES.

The Perch	194	The Bleak	227
The Pope or Ruffe	197	The Minnow	228
The Basse	199	The Twaite Shad	230
The Miller's Thumb	200	The Salmon	235
The Three-spined Stickleback	201	(1) Salmon, (2) Sea-Trout, and (3) Trout Parr	238
The Ten-spined Stickleback	202	A Forty-pounder Salmon "showing himself"	253
The Carp	206		
The Barbel	209	The Sea-Trout	258
The Gudgeon	211	The River-Trout	262
The Loach	214	The Grayling	271
The Tench	215	The Char	275
The Crucian Carp	219	The Pike	280
The Bream	220	The Eel	287
The Roach	222	The Caudal Heart of the Eel	290
The Chub	223		
The Dace	226	The Lamprey	291

LIST OF ILLUSTRATIONS. xvii

MOLLUSCS, CRUSTACEANS, AND ANNELIDÆ.

	PAGE		PAGE
Grey and Black Slugs	293	The Crayfish	297
Fresh-water Snails	294	The Fresh-water Shrimp	298
Helix aspersa	295		

INSECTS.

Gyrinus natator (Whirligigs)	301	*Nemoura variegata*	322
The Great Water-Beetle	302	(1) The Common Grasshopper and (2) The Great Green Grasshopper	322
The Water-Boatman	303		
The Water-Bug	303		
The Devil's Coach-Horse	304	Pupa (nymph) of *Ephemera vulgata*—May fly	330
The Dor-Beetle	306		
Pœderus riparius	307	*Ephemera danica*	332, 333
The Marlow Buzz	307	*Ephemera vulgata*	333, 334
Odocantha melanusa	307	Outline Eyes of Baëtis, magnified	340
Wing of Neuropterous Insect	309	Outline Eyes of Cloëon, magnified	341
The Great Dragon-Fly	310		
The Dragon-Fly or Horse-Singer	311	(1) Baëtis with forewing elongated at apex; (2) Cloëon with forewing rounded at apex	343
The *Libellula puella*	312		
Pupa of the Dragon-Fly	314		
Different Forms of Caddis-Cases with the Silk Grating	316	*Baëtis fluminum*	345
		Baëtis lateralis	345
		The Red Ant	347
Phryganea varia	318	The Saw-Fly	349
The Alder-Fly	319	The Governor	349
Perla marginata	320	The Water-Scorpion	350
The Yellow Sally	321	The Frog-Hopper	350

FLOWERS.

The Sweet Flag	369	The Club or Bull Rush	381
Broad-leafed Pond-Weed	374	Common Hornwort	382
Yellow Water-Iris	377	Seed-vessel of *Nuphar lutea*	385
The Frog-Bit—Fertile flower, sterile flower, and leaf	380	The Common Reed	388
		The Reed Mace	389

THE

ERRATA.

Page 64, line 7, *for* "Willoughby" *read* "Willughby."
,, 92, ,, 13, ,, "seeks" *read* "decks."
,, 117, ,, 3, ,, "*Fringellidæ*" read "*Fringillidæ*."
,, 120, ,, 12, ,, "*cannatina*" read "*cannabina*."
,, 146, ,, 20, ,, "Tous" *read* "Sous."
,, 194, ,, 19, *after* "path" *read* "pursue."
,, 379, ,, 9, *for* "DOG-BEAN" *read* "BOG-BEAN."
,, 379, ,, 10, ,, "*trifoliatum*" read "*trifoliata*."
,, 381, ,, 11, ,, "*Scripus*" read "*Scirpus*."

---

the rat is of the family *Muridæ*. The vole was formerly considered to be an aberrant form of the beaver family, but through the researches of Mr. Waterhouse it is now placed between the American musk-rat and the lemmings.

This interesting little animal is, we believe, a pure vegetarian, although Professor Huxley (The Crayfish, p. 9), states that it is extremely partial to this crustacean as an

A

THE RIVER-SIDE NATURALIST.

CHAPTER I.

THE QUADRUPEDS.

THE WATER-VOLE.

AMONGST the quadrupeds the first we shall notice is what is commonly called the WATER-RAT, but which is no rat at all, as it belongs to a totally different family, although of the same natural order (the rodents). Its proper designation is the WATER-VOLE, of the family *Arvicolidæ*, whilst the rat is of the family *Muridæ*. The vole was formerly considered to be an aberrant form of the beaver family, but through the researches of Mr. Waterhouse it is now placed between the American musk-rat and the lemmings.

This interesting little animal is, we believe, a pure vegetarian, although Professor Huxley (The Crayfish, p. 9), states that it is extremely partial to this crustacean as an article of diet. "It is averred," he says, "that the water-rat is liable to be seized and devoured. Passing too near the fatal den, possibly in search of a stray crayfish, whose flavour he highly appreciates, the vole is himself seized and held till he is suffocated, when his captor easily reverses the conditions of the anticipated meal." Others assert that the vole will eat eels, young fish, young ducks, and the like. All these assertions arise from this animal being mistaken for the common brown rat (*Mus decumanus*), which in summer and autumn deserts the homesteads and houses, and takes up its abode in the hedgerows and by the water-side, and will devour all kinds of animal matter dead or alive, and, as a writer in *Household Words* remarks, "making sad havoc amongst the fish that come wandering by," and thus the water-vole gets the blame. This similarity to the brown rat is most unfortunate for the poor beast, for besides its natural enemies, the weasel, the stoat, the owl, &c., it is hunted to death by the river-keepers, who declare that it eats the spawn of trout and other fish. Isaac Walton, so kind and so gentle to all living creatures, places the craber, which some call the water-rat, by the side of the otter and cormorant, although he declines to "quarrel with it, as he loves to kill nothing but fish." A correspondent in the *Fishing Gazette*, March 1887, says: "During a long experience about the water-meadows at all seasons, in the character of angler, naturalist, or sportsman, I have never come upon the vole eating fish;" and as far as we know, there is no record of any one detecting it eating anything but vegetable food.

The water-vole is easily distinguished from the common rat. The shape of the head is rounder, the ears much smaller, the tail covered with thick reddish-brown hairs and comparatively short (see Fig. 1); whereas in the rat the head is more pointed, the eyes more prominent, the ears much longer, and the tail is naked, ringed, and scaly, with a few fine hairs, and considerably more lengthened (see Fig. 2).

The water-vole is also much more red in colour, almost uniform all over. The common rat has whitish under parts.

THE WATER-VOLE.

The teeth also are different in the water-vole, and there is a slight inclination to a web in the hind feet. The whole length, including the tail, is about 13 inches, while the whole length of the common rat is $16\frac{1}{2}$ inches.

The water-vole has been seen at times to carry its young across a river in its mouth. Mr. G. T. Rope, writing in the *Zoologist*, says: "Walking by the side of a stream early in May, I saw a large water-rat carrying in its mouth a half-grown young one. While swimming the young rat was held well up out of the water; sometimes, however, the old one would leave the water and cross a bit of mud, still holding the youngster, which while in sight it never once dropped.

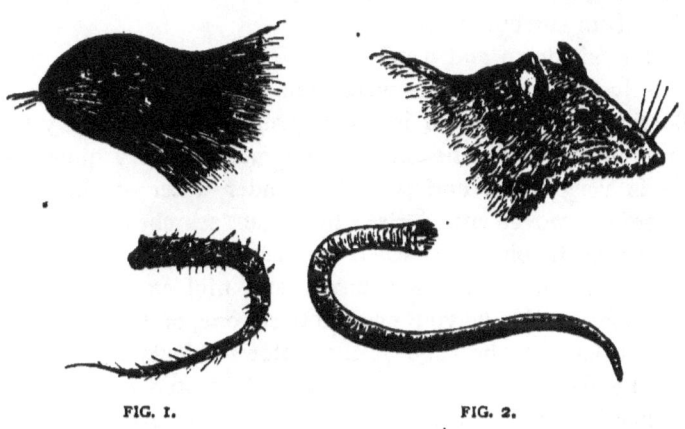

FIG. 1.   FIG. 2.

It presently, however, disappeared round a bend, but in a few minutes I again saw it returning without its burden, which had doubtless been lodged in a place of safety. She seemed in a great hurry, and was perhaps going back for another young one. Probably the nest had in some way been disturbed, or was threatened with danger, and the family were removing to safer quarters. It seems strange, however, that the youngsters, which, judging by the size of the one I saw, were certainly pretty well half-grown, could not be trusted to follow their dam without help. The removal of the whole family by this means must have been no slight undertaking, as the distance she carried the young one, while in my sight, was at a rough guess from twenty to thirty yards, and she may have carried it some way before I noticed her. Probably there were from four to six of them to move, that being, I believe, about the average number in a litter. In April 1871 I found three nests, in one of which the young were no bigger than full-grown house mice, but were covered with hair and could swim and dive well. The glossy fur of this little animal receives reflections very readily, and is, I think, a means provided for the safety of its possessor. Many persons must have noticed how difficult it is to catch sight of a water-rat sitting perfectly still on wet mud or in some similar situation, by reason of the fur receiving by reflection the general tint of its surroundings."

Just sit down and watch for a few moments. See how quietly and suddenly the water-vole appears on that mass of floating weed, retarded in its progress down-stream by the growing reeds at the side. You must be very quiet, for he is very wary, and will dive under water at the least noise or movement. See how he searches for some particular tit-bit, and when he has found it he sits up on his haunches like a squirrel, and nibbles away till it is finished. Have you ever noticed one of those sword-leaved flags at the edge of the water suddenly topple over and lie flat on the surface? It is the water-vole's work; he has bitten it through at the bottom, and now rises to the surface to enjoy his repast on the sweet, succulent root.

How often, when silently walking up-stream, one is startled by a sudden flop in the water, and then a water-vole rises half-way across the river, making for the opposite bank; or you may watch him quietly swimming close in-shore to find refuge in his burrow. This pretty, harmless animal feeds almost entirely on water-plants. It is particularly partial to the spongy roots of the different species of flags, more especially the common sweet flag (*Acorus calamus*), and in the rivers and streams where this plant grows the water-vole abounds.

A dark variety, almost black, of the common water-vole is sometimes met with.

Two smaller species of this family frequent the adjacent meadows; the pretty little SHORT-TAILED FIELD-MOUSE (*Arvicola agrestis*), found in damp meadows and other places, often locally called the water-mouse. Its usual food consists of herbs and roots, but it is said it will eat flesh when pressed for hunger; but it is not so fond of insect or animal food as the RED FIELD-VOLE (*Arvicola glareolus*), commonly known as the bank-vole, which is easily distinguished by its brighter red colour, more glossy fur, and the length of the tail, besides the peculiarity of the teeth. Bell says: "This most remarkable peculiarity is the development in the adult of distinct roots to the molar teeth; the first upper grinder has four cemental spaces and five angles, the second has six spaces and eight angles, and the first lower grinder has seven spaces and nine angles."

These pretty little animals seldom come under observation unless specially sought after.

THE OTTER.

A few years ago otters were rarely found farther south than our midland counties. At the present time there is scarcely a river in the United Kingdom in which this animal is not trapped. In our southern chalk-streams otters have become rather formidable in their numbers, and yet these rivers are perhaps more carefully watched than any other; but otters are great wanderers. It is more than probable that the strict preservation of fish of all kinds, particularly of the Salmonidæ, is the cause of this great increase.

The OTTER (*Lutra vulgaris*) is of the family *Mustelidæ*. The body is elongated and low. The feet have five toes, all palmated. The tail is flattened horizontally, the tongue roughish, and the ears small.

The otter appears to have been considered, if not a fish, certainly very fishy by the Church of Rome, and was allowed to be eaten on fast-days. Pennant saw one being cooked for dinner in the Carthusian Convent at Dijon. This religious order cannot by their rules ever partake of flesh. Walton makes Venator say, when he asks him if he hunts a beast or a fish, "Sir, it is not in my power to resolve you—I leave it to be resolved by the College of Carthusians, who have made vows never to eat flesh. But I have heard the question hath been debated among many great clubs, and they seem to differ about it; yet most agree that her tail is fish, and if her body be fish too, then I may say that a fish will walk upon land (for an otter does so) sometimes five or six, or ten, miles in a night."

THE OTTER.

Although for the most part a fish-eater, there is considerable difference of opinion as to the advisability of destroying the otter in trout-streams — whether its particular predilection for eels and small jack does not compensate for any harm it may do to the trout, as there can be but little doubt that eels destroy an enormous quantity of trout ova and fry.

Mr. Collier, a Master of Otter Hounds, states that the otter is in reality the trout-angler's friend, from being the

deadliest foe to the eel, which is in turn the deadliest
enemy to the trout-angler, as eels will prove more harmful
to a trout-stream by destroying the spawn than the otters
will do by killing the fish; in fact, Mr. Collier says if
he owned a trout-stream he would never allow an otter
hound on it. An "Old Friend of the Otter" writes:—
"I have opened forty-five hunted otters, and many of them
killed before six o'clock in the morning, and only found in
two any parts of fish about three inches long, and when
their stomachs were cleaned in running water I found it
was eel; and I feel sure their food is the eel; but in the
spring of the year eels being deep in the mud, they may
catch an occasional kelt and suck a little of the blood out
of its shoulders. I would say to any gentleman, If you
wish trout to be plentiful, preserve the otter and he
will eat the eel, who lives upon fish-spawn, especially the
trout's."

The following is taken from the *Field* of May 7, 1887 :—
"In his report on the upper waters of the Severn, made to
the Fishery Conservators last week, Mr. George has brought
a most serious charge against the otters which frequent that
district, in the statement that, of 250 salmon found dead,
ninety-nine had been killed by otters, thirty-six by disease,
and the cause of death of the remainder could not be
ascertained. For many years I observed and studied the
habits of the otter in confinement as well as in its natural
state, and my opportunities for so doing were unusually
favourable, living, as I have done for a quarter of a century,
by the banks of a river where otters were far from a rarity,
and fish numerous enough. During the whole of that
time, a great portion of which was spent by the river-side,
I do not remember to have seen more than one salmon
whose death could actually be ascribed to an otter, and the
remains of a trout similarly destroyed I never observed.
Relics of roach, that when in life had become the prey of
the otter, were common enough; portions of eels, which
had no doubt been similarly treated, were occasionally
found; and the droppings of the otter, as a rule, contained
the bones and scales of small fish, the shells of the river

crayfish, and the feathers of water-fowls, water-hens for the most part. Then, whilst otter-hunting with the local pack of hounds, the line of scent was not infrequently carried from the river across the meadows on to some swampy ground and rushy bottom, which no doubt our quarry had visited for the purpose of making a raid upon the frogs; but, personally, I never saw the remains of these creatures in the coke of the animal, though others with whom I am acquainted have often done so. Sundry otters that I have known, when kept in confinement, have proved by no means fond of salmon, far preferring flounders, eels, and crayfish to the more valuable fish. The latter they were often tempted to take by way of experiment, and showed no great liking thereto. Indeed, Troughton, the huntsman to the Kendal hounds, told me that one otter he kept in captivity enjoyed a young rabbit for a meal as much as anything.

"Can these otters in the Severn Mr. George writes about feed differently from those inhabiting that district with which the writer is best acquainted ? Severn salmon, I am aware, are particularly rich in curd, and choice in flavour and excellence. The otters may know this, and feed upon them accordingly; still, it would be interesting to all hunters of the otter to learn by what means or appearance it was so decisively known that these ninety-nine salmon had been killed by otters. The little tit-bit eaten out of the shoulders we have all heard of, but I have seen a common rat engaged on a dead salmon picking out that identical portion which is said to be the *bonne bouche* of the animal of which I write, and the carrion crow and the magpie I have seen pecking away at the shoulders of a dead salmon lying on the water's brink. The poor otter has many enemies, and, alas! few friends, but I must say that those persons who are best acquainted with his habits are the very ones who come forward in his defence; and such is not usually the case unless the animal attacked is more sinned against than sinning. That the otter is the latter I have no manner of doubt."

Wonderful stories are told of otters dragging to shore

large salmon of 20 lbs. and upwards, and then only nibbling a little bit out of the back of the neck. These may be believed or not, according to the imagination of the reader. We have never been able to authenticate in a single instance any one having seen an otter perform such a feat.

In the winter, when ice and snow has driven them from their usual haunts, otters, when thus pressed for food, will frequent the neighbourhood of cottages and homesteads, and will eat anything they can pick up, from a duck to a rat, and will not despise a rabbit should that rodent come in its way. A writer in the *Fishing Gazette*, hailing from the north, says the greater part of the diet of the otter consists of crayfish, thousands of which it destroys, and it is for these that long journeys are so frequently made.

In hunting the otter certain peculiar terms are always employed. The foot-mark is "the seal," the dropping "the spraint," and when the animal rises to the surface to breathe, "to vent."

The length of a full-grown otter is from 3 feet to 3 feet 6 inches, including the tail; colour brown, with the throat, cheeks, and under parts of a whitish-grey.

The fur of the otter is of two kinds—the outer with long coarse hairs, the inner fine and soft.

The average weight of a full-grown otter, according to Bell ("British Quadrupeds," second edition), is from 15 lbs. to 18 lbs. Some are occasionally killed weighing over 20 lbs.—one recorded in the *Field*, February 20, 1886, of the enormous weight of 36 lbs. I much doubt if the biggest of these otters could chase, seize, and bring on shore a salmon of 15 lbs. or 20 lbs., or even of much less weight. But we know that two-legged otters, much heavier, are apt to frequent salmon pools and streams, at all times and all seasons, and they have nasty jagged teeth, although not in their mouths.

This animal, from his nocturnal habits, rarely comes under the observation of the angler. We had an opportunity last year of watching one fishing on the Kennet. We had strolled up late in the evening to a certain pool, to see if any of the big fish were rising, and we were standing very still under the shadow of an old pollard willow-tree, when we saw something swimming across the river, making a considerable wave, to the opposite bank. It was bright moonlight, and we saw at once it was the otter (we knew he frequented these parts); he half got out amongst the reeds, and then with a silent plunge dived again. When he came up he was close on our side, and he immediately spotted us, threw up his head for a second to be certain, and then at once dived and was away. Some naturalists assert that the otter, from the peculiar position of his eyes, cannot take a fish except in deep water, because he must be *under* the fish before he can see him; but we have found the remains of trout on the bank of very shallow streams a long way from any deep water.

THE WEASEL.

Of the same family as the otter (*Mustelidæ*), that very restless and lively little animal the WEASEL (*Mustela vulgaris*) is very frequently seen on the river-banks, especially where there are bushes, old stumps of willows and alders, running in and out amongst the roots, climbing along the bushes, now half-way up the stem of an old willow, now disappearing in a rabbit-hole or in the thick vegetation, hard at work searching for its prey, which chiefly consists of rats and mice, with occasionally a young rabbit. Not that it is at all squeamish if it comes across a nest of eggs or of young birds, having a particular propensity to suck eggs and swallow tender morsels.

THE WEASEL.

This little animal does more good than harm to the farmer by frequenting the stackyard and destroying a great number of rats and mice in the corn-stacks. At times, when driven by hunger, he preys upon young chickens and pheasants, but its regular food is of the more ignoble kind. It has also been stated that the weasel, like the mungoose, attacks and eats snakes, but from experiments made to verify this it has been found not to be the case.

The weasel generally kills its prey, if a mouse or a rat, by biting it on the head and thus penetrating the brain with its sharp teeth, and it always begins its meals by devouring the brain, not by sucking the blood, as is generally supposed. Owing to its long and supple body, the weasel can pursue its prey with great facility under ground, and it climbs trees with great ease.

The weasel swims easily and rapidly, will cross rivers and streams after its prey, and has been known to carry its young across a river when disturbed from its usual haunts. The following account of a weasel swimming is taken from the *Zoologist* of August 1884:—

"Walking along the river 'wall' near here on the 24th June last, I saw a short distance ahead a strange-looking object swimming across the river to the opposite side, which on landing proved to be a weasel, carrying in its mouth a young one, to all appearance more than half the size of its parent. On landing she found herself suddenly face to face with two colts, upon which she dropped the youngster and ran into a clump of brambles and nettles close at hand, but almost immediately returned; and again taking up the young one, she went 'looping' along through the long grass at a pace which, considering the weight of her burden and the shortness of her legs, was really wonderful. I could not see what ultimately became of her, but at the time I lost sight of her she was apparently making for a tall thick hedge bordering a ditch, where perhaps she had already fixed upon some safe retreat for her family. I have more than once seen a stoat swimming —probably a matter of common occurrence with that species, which is very partial to the banks of rivers, watercourses, ditches, &c., where it preys upon the rats, water-rats, meadow mice (*A. agrestis*), young water-hens, &c., to be found in such places. Only last spring I saw one cross a small stream carrying some object which I took to be a large meadow mouse, but was not near enough to be quite certain. The weasel in this district is much scarcer than the stoat, but neither can by any means be called common, being everywhere persecuted with the utmost rigour by gamekeepers; besides which, the objectionable practice of destroying rats and mice by means of poison, which has become so prevalent of late years, must be very fatal to both species, not to mention the hedgehog, the poisoned rats and mice being in all probability devoured by all three. It is a great pity that the pretty and very useful little 'mouse-hunt' should be so dealt with.—G. T. ROPE (Blaxhall, Suffolk)."

The weasel has a long, lithe body, colour reddish-brown, and white beneath; it lives principally on rats and mice, and from its suppleness is able to follow them in all their runs and holes, in barns or corn-stacks. It is, however, persecuted by the gamekeeper, and is always a conspicuous figure on the "gallows-tree." The length of the male is from nine to ten inches including the tail; the female is much smaller, not much bigger than a large field-mouse. In Hampshire the female, from its small size, was, in Gilbert White's time, considered to be another species, and went by the provincial name of cane, or kine.

The weasel is easily distinguished from the stoat by its smaller size, and having *no black tip to its tail*.

THE STOAT.

Of the same family as the weasel and otter. The STOAT or ERMINE WEASEL (*Mustela erminia*) is not so often seen by the river-side, but being very partial to the young water-vole as an article of its diet, it at times comes down to the river-bank and will often swim the stream in pursuit of its prey. It is considerably larger than the weasel, but

THE STOAT.

quite as active. It is especially fond of rabbits, and hence keeps to the thickets and other places where these animals abound. You can tell it at once by the black-tipped tail. In summer its colour is reddish-brown, with white belly and throat; in winter brown and white, and often quite white—it then becomes the ermine of commerce. When hunting rabbits it appears as if it fascinated them after a short time, for the rabbit will suddenly stop and utter a most piteous cry, and if nothing intervenes the victim is soon silenced. We have often seen this, and rescued many a rabbit; but it is most difficult, almost impossible, except by death, to drive the stoat from its victim. Although frightened at first by your presence, it will return again and again to take up the scent. The stoat is a most

THE STOAT.

courageous little beast, and will defend its young against all odds. We once came across a male and female with four young ones migrating from one part of Richmond Park to another. On approaching them, both the old ones set up a defiant chatter and rushed towards us, discontinuing their attacks only with death.

Very few stoats in their white winter dress are seen in our southern counties, but in the alpine districts of Wales and Scotland and Northumberland it is very common to find them; indeed in Scotland and Wales the change of colour is almost universal. The skins obtained in this country are very inferior, both in beauty and value, to those from Russia, Norway, and Lapland.

Bell ("British Quadrupeds," second edition) says the definition of the word *stoat* is very probably from the Belgic

"*Stout*," bold; and the name is thus pronounced at the present time in Cambridgeshire and elsewhere. The stoat is one-third larger than the weasel; the head is broader and the tail longer, the tip being black and rather bushy.

The muscles of the neck of both stoat and weasel are extremely powerful and well developed, as the following anecdote—witnessed by a friend whose veracity we can vouch for—will prove. One evening he noticed, when in Richmond Park, two stoats dragging a dead rabbit up the inclined stem of an old pollard oak, and disappear into a hole in the trunk. On tapping the tree with a stick the two old stoats rushed out with a most prodigious chattering, and immediately afterwards four or five young ones. A young lad who accompanied him climbed the tree, and put his hand and arm down into the hole, when, to our friend's astonishment, he pulled out a rabbit, dead, but quite warm, and the remains of four others.

THE SHREW-MOUSE.

Many of us, no doubt, in our rambles, have come across, lying dead on the gravel-walks, a mouselike-looking animal with a long snout. This is the SHREW-MOUSE (*Sorex vulgaris*, Order *Insectivoræ*, Family *Soricidæ*).

THE SHREW-MOUSE.

Although called a mouse by common consent, it is nothing of the kind, but is a *sorex*, and its only likeness to a mouse is in its colour. Many suggestions have been started as to the cause of its death. Some attribute it to cats and owls mistaking it for the common mouse, and not eating it on account of its peculiar smell. This may perhaps apply to the cat, but not to the owl, as the bones and skulls of the shrew-mice have been constantly found in the owl's castings. The shrew-mouse is a most pugnacious little beast; may not the death be caused by a pitched battle between two of them? It appears that these little animals are mostly found dead in the autumn. Bell says: "The cause does not appear to be understood. So many may be found at this season lying dead on footways or on bare ground near their haunts, as to have led to the belief among country people that the shrew could not cross a public way without incurring instant death."

A strange superstition formerly existed in regard to this little harmless animal. It was supposed to be able to inflict very great pain to cattle by running over them at night when lying down, and the only specific against this disease was whipping the afflicted beast with a branch from what was called a shrew-ash. This particular tree was to be found in many old villages. Gilbert White mentions one on the Plestor at Selborne. The tree was made by boring a hole in the trunk of a pollard or other ash, and inserting into this hole a live shrew-mouse, and then plugging the hole with a piece of the same tree. When the mouse was supposed to be dead and its juices had entered into the sap of the tree, then the branches were fit for use. Plott, in his "History of Staffordshire," states that some workmen sawing a trunk of solid oak cut through the body of a hardishrew, or nursrow, as they call them, *i.e.*, a field-mouse, so that a shrew or nursrow tree was not confined to the ash.

In Broderip's "Zoological Recreations," p. 91, we find the following :—

"The common shrew-mouse, one of the most harmless of animals, was considered to be a very pernicious creature. Its bite was held to be venomous by the ancients, and our own ancestors believed that if a shrew-mouse ran over the limbs of man or beast, paralysis of those limbs was the consequence; hence, perhaps, the old malediction, "*Beshrew thee!*"

The shrew is about $2\frac{1}{2}$ inches in length from the snout to the root of the tail—the tail about 1 inch 7 lines. It is found in dry fields, orchards, gardens, and hedge-banks, feeds chiefly on insects and worms, for which it pokes its long snout into the dead leaves and roots. It burrows or makes long runs just under the surface of the ground.

The colour is reddish mouse-colour above, grey below. In some the brown on the back is very dark, in others a light chestnut, ears very small and rounded, snout long and thin, body short, back somewhat elevated, tail shorter than the body.

Shrews are difficult to see when alive, as they are very shy and keep much to the herbage and amongst the dead leaves. They are very fond of the slug (*Limax agrestis*), as well as of insects and worms.

THE WATER-SHREW.

There is another small shrew which frequents the rivers and streams, the WATER-SHREW (*Sorex fodiens*), which we may place amongst the aquatic animals, as he finds his food almost entirely in and on the water. Shrews are, for the most part, insect-feeders; but this little fellow, we are afraid, is otherwise disposed, and much doubt if he is not very fond of the ova of trout and other fish. Bell ("British Quadrupeds," second edition) says that this animal may be often seen at the bottom of streams and ditches, turning over the stones. "The food appears to be taken at the moment the stone was raised from its resting-place, though in some instances by the animal merely poking its long snout under the stone without lifting it, but in every case when caught it was conveyed to the side to be devoured. . . . This food appears to be chiefly composed of the *Gammarus pulex*, a sessile-eyed crustacean inhabiting our streams." Bell goes on to say: "We do not know whether the water-shrew is piscivorous in its habits, though it is not unlikely that it may feed on the spawn or fry of minnows and other small fish; but to its carnivorous propensities we can ourselves bear testimony."

This water-shrew is a very pretty little animal, nearly black over its back and upper parts, with perfect white on belly and under parts. It is very rapid in its motion when under water, dives and swims with the greatest velocity, and when under water its coat is covered with bubbles of air, looking like silver globules, the hair being perfectly impermeable to water. It is a favourite food of the weasel, who will often follow it into the water; but as the weasel is not an adept at diving, Master Shrew beats him in this element, and thus saves his life.

There is another species (*Sorex pygmans*), the lesser shrew, found throughout England, Scotland, and Ireland. Its habits are similar to the common shrew, but it is much smaller. It is subject to the same mysterious mortality in the autumn.

THE BATS.

Some of our bats are constantly found by the river-side, at eventide, particularly the GREAT BAT (*Scotophilus noctula*), the PIPISTRELLE (*S. pipistrellus*), DAUBENTON's BAT (*Vespertilio Daubentonii*), and the LONG-EARED BAT (*Plecotus auritus*), all of the family *Vespertilionidæ*.

Much superstition was formerly attached to bats, probably on account of the earlier naturalists believing them to be neither fowls of the air nor beasts of the field. Aristotle

B

speaks of them as birds with wings of skin. Pliny says they were birds which brought forth their young alive and suckled them. So late as the time of Buffon this ignorance of their real nature existed. He says: "An animal which, like the bat, is half quadruped and half bird, and which in fact is neither one nor the other, is a kind of monster." Their appearance only in the evening and at night has connected them with deeds of darkness, with witches, and even with the king of evil himself. "Wool of bat" is one of the ingredients of the witches' charm, and even at the present time some have a passing shudder when a bat flies across their path. It is strange to think that our little harmless bats, "whose habits," says Bell, "are at once so innocent and so amusing, and whose time of appearance and activity is that when everything around would lead the mind to tranquillity and peace," should be so mixed up with superstition and mystery.

We may often notice late of a summer's eve a large bat flying high in the air, making long circuits, but generally reappearing after a few minutes; this is the GREAT BAT or NOCTULE (*Scotophilus noctula*), one of the largest of our bats. Its food consists chiefly of coleopterous insects—*i.e.*, beetles and the like—and it is particularly partial to the cockchafer (*Melolontha vulgaris*). This bat flies with great rapidity, and if watched for a few moments, a very peculiar motion may be observed during the flight. It is, as Bell describes it, like the fall of a tumbler-pigeon, and is produced by the animal closing its wings for a moment or two whilst it uses its armed thumb to transfix some big beetle which it is unable to swallow at one gulp, so that it may devour it more easily. The length of the body is nearly 3 inches, and the expanse of the wings from 13 inches to 14 inches, sometimes even greater.

The most common bat we see is much smaller than the above—viz., the PIPISTRELLE, or Flitter-Mouse (*Scotophilus pipistrellus*). It is one of the least, as the noctule is one of the largest, of our bats. This little fellow comes out much earlier—often, indeed, in dark gloomy weather long before evening—and flitters around us, seeking out the places where insects most do congregate, now round the barn or haystack, and now along the river-side, contending almost with the trout for the sedge and other flies. When you look at it closely you will see that it has a russet-grey fur, dark leathery wings, and very wee eyes. There is another bat not so common, but often confounded with the pipistrelle, namely, DAUBENTON'S BAT (*Vespertilio Daubentonii*), which frequents the surface of the water more especially, and has a much slower and more quivering flight. Bell says it is essentially an aquatic species. The difference is in the expansion of the wings, Daubenton's being 9 inches, the pipistrelle 8 inches 4 lines. The colour of fur is a reddish-brown, a moustache of soft, long hair on each side of upper lip, and the foot free from the wing membrane; while in the pipistrelle the colour is yellowish-red on forehead, a protuberance on each side of nose, and a small elevated wart over each eye.

Another small bat is known as the LONG-EARED BAT (*Plecotus auritus*), distinguished principally from the extraordinary length of its ears in comparison to the rest of the body; not so often seen as the pipistrelle, but common enough in some parts; is somewhat larger than the flittermouse, and it is difficult to know one from the other in the flight; but the long-eared bat has a very peculiar voice, a kind of a shrill chatter, which, when once heard, will readily distinguish it. When in hand it is at once recognised by the extraordinarily long, transparent ears. These bats become very tame and much attached to those who feed them, often taking a fly from the lips.

Another not very uncommon bat we may mention is the BARBASTELLE (*Barbastellus Daubentonii*). Bell says: "If in a twilight stroll about midsummer a person finds himself in close proximity with a bat of somewhat thick and clumsy form, but of rather small size, whose flight is so desultory that it appears to be flapping lazily about hither and thither, seemingly without purpose, and intruding so closely that the flutter of its wings may be heard, and even the cool air thrown by their movement felt upon the cheek, it may with almost certainty be recognised as the barbastelle."

All our bats are nocturnal in their habits, although one or two of them often appear in the gloaming, and at times even at midday, but as a rule the day is passed in sleep in the darkest places they can find, in hollows of trees, old ruins, towers of churches, dark barns, and the like. They are all insectivorous and fly with remarkable rapidity. The ease with which they turn and twist about in their flight and in pursuing their prey is extremely interesting to watch on a calm summer's eve. The skin of the wing is so sensitive that it was found by Spallanzani that if bats are deprived of their sight and hearing, they were able to fly about with absolute certainty, and avoid objects purposely placed in their way. The ears of bats, forming so important a part in their economy, are therefore much larger than would appear necessary for the size of the animal.

HEAD OF THE GREAT BAT.

CHAPTER II.

THE BIRDS.

BIRDS have been associated with man from time immemorial, and have been chosen as a favourite theme of song by the poets of all ages and of every country. Alike in Pagan as in Christian times, whenever the virtues or the vices, the loves or the hatreds, the victories or the defeats, of men have been sung, the birds rarely fail to be alluded to either directly, or in simile, or metaphor.

Birds in ancient times, either by their flight or their sudden appearance, often decided the destinies of nations, the march of armies, the fall of cities, or the reigns of sovereigns. The flight of birds was anxiously regarded by the augurs, and interpreted by them as indication of success or defeat, of peace or of war. The cackling of geese saved Rome, and caused these birds, so despised in our days, to be held in veneration and esteem.

Birds held a prominent place in mythological history. The eagle was the bird of Jove; Juno had her peacocks, Minerva her owl, Venus her doves and sparrows. Fables of men and women transformed into birds and animals abound in the poetry of the ancients, and Ovid in his "Metamorphoses" only repeats and enlarges on the all-prevailing superstition of those early times.

The legends of the transformation of Philomela into a *nightingale*, Procne into a *swallow*, Tereus into a *lapwing*, Antigone into a *stork*, Alcyone into a *kingfisher*, Cygnus into a *swan*, only prove how largely the feathered races entered into the imaginations of the poets of those days.

In later times birds still held their place in poetical literature. Heroes are likened to the eagle and falcon, cravens to the kite and the crow. High-born dames and lovely damsels of the days of chivalry were compared by the minstrels to some fair feathered denizens of the air or woods. In time of war the noble or the knight placed the effigy of an eagle or a falcon on his helmet in token of his high bearing and courage. In time of peace he held the living falcon on his gloved wrist, while his dame was accompanied by the merlin or the hobby.

In Guillim's "English Heraldry" birds and parts of birds enter largely as heraldic devices. Eagles, eagles' heads, falcons, swans, cygnets, wild ducks, geese, sea-mews, shovellers, cormorants, storks, kingfishers, owls, ravens, pelicans, pheasants, bustards, choughs, swallows, martletts, turtle-doves, and many others were used as crests and supporters to arms.

Birds figure as crests in many a noble family. The Belmores have a cock, the Boileaus a pelican, the Bridports a chough, the Cawdors a swan, the Cannings a cormorant, Chelmsford a dove, Denbigh a nuthatch, Derby an eagle, Falmouth a falcon, Galway a swallow, Leitrim a lark, Lytton a bittern, Mexborough an owl, Strickland a turkey, Temple a martin, and so on through a long list.

THE KINGFISHER.

One of the most beautiful of our river-side birds is the KINGFISHER (*Alcedo ispida*). Its rapid flight as it passes up or down the stream, "swift as a meteor's shooting flame," prevents a close inspection; but sometimes, when the bird is perched on some bare branch projecting over the water intently watching for its prey, a chance may occur. It will then be seen with what vivid colours it is clothed. The crown of the head and nape of the neck very dark green, with bars of glossy blue-green; the wings dark green, upper wing-coverts turquoise blue, back and tail glossy greenish-blue; chin and throat light buff, with two bands of blue and green extending from the base of the bill to the side of the breast; under parts chestnut, legs reddish.

THE KINGFISHER.

This bird is becoming scarcer every year; at one time it was persecuted for the sake of its beautiful plumage to adorn a lady's bonnet; but now another and very formidable enemy has entered the field. The trout-breeders have declared war to the knife against this poor bird, and nothing will satisfy them but its complete extermination.

We are quite aware that it does destroy a large number of small fish—sticklebacks, minnows, small trout, and the

THE KINGFISHER.

like; but fish do not comprise the whole of his diet. It does an immense amount of good by destroying vast numbers of the greatest enemies to the young trout. Yarrell (4th edition) says: "Its food consists of small crustaceans, aquatic insects—as dragon-flies and water-beetles," the larvæ of which are very destructive to trout fry, "and little fishes, especially minnows and sticklebacks, both of which prey upon the ova and fry of other fish, whilst leeches are also said to enter into its diet."

One mode of capturing these birds, as practised in many places, is extremely cruel. A number of small spring-traps are set on narrow boards crossing the stream, on the hatches, and on old boughs hanging over the river, on which kingfishers often perch, and when settling on these the trap is sprung, and catches the poor birds across the legs, breaking and tearing them. There they remain, may be for many hours, till the river-keeper visits his traps. It often happens that the trap cuts the legs clean off, and the poor maimed bird flies away to die a lingering and terrible death by starvation.

In the Highlands of Scotland, where the proprietors or the tenants prevent the destruction of the golden eagle or the peregrine falcon, the grouse on these moors are found to be just as plentiful, and much more healthy, as on those where the birds of prey are ruthlessly exterminated. In the same way, the kingfishers cannot hurt a trout-river. Nothing is more detrimental than overstocking—it breeds disease in the grouse on the moor, and lanky, unhealthy fish in the river.

The kingfisher has a most interesting mythological history. Aristotle, who died 320 B.C., writes of its powers of calming the winds when sitting on its eggs in the sea-girt nest. Look into Ovid or Lemprière, and read how Alcyone, daughter of Œolus, married Ceyx, who, unfortunately, was drowned in a great storm; and when Alcyone found his dead body on the shore, she threw herself into the sea, and was changed, together with her husband, into kingfishers, with the permission to keep the waters calm and serene for the space of seven to fourteen days whilst they built their nest on the surface of the ocean. Hence "calm" days and "halcyon" days are synonymous terms:—

> "There came the halcyon, whom the sea obeys,
> When she her nest upon the water lays."

Thus wrote Drayton, and thus the poets write up to the present time. Keats says:—

> "O magic sleep; O comfortable bird
> That broodest o'er the troubled sea of wind
> Till it is hush'd and smooth."

The French call the kingfisher "Martin Pecheur," in allusion to St. Martin's summer; and Shakespeare says :—

"Expect Saint Martin's summer; halycon days."

There is a curious notion prevalent in some of the counties of Great Britain that when the skin of the kingfisher is hung up by a thread, the beak will always turn to the quarter the wind is blowing from, or even before a storm commences, and this curious weather-gauge may still be found in many cottages. Shakespeare alludes to this in "King Lear:"—

"Revenge affirm, and turn their halcyon beaks
With every gale."

Christopher Marlowe also, in the "Jew of Malta:"—

"Into what corner peers my halcyon bill?
Ha! to the east."

We plead for the kingfisher. Let us hope more merciful and more sensible councils will prevail, and that we may all again be delighted to watch the bright hues and rapid flight of this "gem of the waters."

THE SEDGE-WARBLER.

When by the river-side we are often little aware how we are surrounded by animal life, how many pairs of eyes are intently watching our movements. One bird in particular is a constant companion, always hiding when one is on the move; but if for a moment one remains perfectly still, a short babbling song, a little harsh in its note, issues from the reeds, and a pretty little brown bird will probably appear moving up the stems or resting on the top of a flag, and will pour forth a series of varied notes, often imitating those of other birds, at the same time being quite aware of your presence. Any quick or hurried movement and he at once disappears. His babble will cease, and he will utter his low, warning call-note—*Churr-churr-churr;* but

THE SEDGE-WARBLER.

he is a fearless little fellow, and is not a bit afraid. Just cast a stone or a clod of earth to where you last saw him, and he will at once resent such impertinence by bursting out again in full song, though in somewhat a lower key. This amusing bird is the SEDGE-WARBLER (*Acrocephalus schœnobœnus*), two crackjaw words, which mean, in plain English, "the pointed-headed bird of the sedges and bulrushes." Of the family *Sylviidæ;* it is commonly known on the banks of the Thames as the Chat. In Ireland it is called the Irish nightingale, as it often sings through the night. It is a summer resident, coming in April and departing in October. On its arrival it at once takes up its abode among the reeds and flags, and builds its nest with grass and bents, placed low down; lays four or five spotted eggs of a yellowish red-brown colour. Seebohm ("British Birds and their Eggs") says "that its haunts are as much in the tangled brake and dense vegetation of marshy plantations as amongst the ever-murmuring reeds." The bird itself is from 4 inches to 5 inches in length, of a rufous brown, streaked with darker brown. A broad streak of yellowish white extends from the beak back over the eyes and ear-coverts. The breast and lower part is of a pale buff, and notice the broad white streak over the eye. One can easily see these marks by using the binocular, and in this manner can distinguish this bird from another which frequents the reeds, but which is much more shy, namely—

THE REED-WARBLER.

The REED-WARBLER (*Acrocephalus streperus*); that is, the Long-Headed, Noisy, or Bustling Bird.

Seebohm calls it *arundinaceous*, or frequenter of reeds, which is much more characteristic. It is of the same order and family as the sedge-warbler, but is larger and more slender, and, from its peculiarly shy habits, not so often seen or heard, although it is probably quite as common. If you are quiet, and in the vicinity of the tall reeds (*Arundo phragmites*), one occasionally is seen flying over the drooping panicles. Its song is often mistaken for that of the sedge-warbler, but the notes are much sweeter. Mr. Dresser ("Birds of Europe"), quoting from Naumann, gives the notes thus: *Tiri-tiri-tiri, tier-tier-tier-zach-zach-zach, zeri-zeri-zeri, tiri-tiri-scherch, scherch-scherch, heid-heid, heid, tret-tret-tret,* and says the entire song is rather a babbling melody than a song. The bird often sings at night, more especially in calm, close weather.

THE REED-WARBLER.

The colour is a pale uniform rufous olive, with a reddish tinge above the tail, a very pale yellow streak over the eye, but nothing like so distinct as that of the sedge-warbler; breast and under parts pale yellow buff. It builds a beautiful nest interlaced in the stems of reeds, generally about half-way up, and is very deep, so that it is not disturbed by the wind. The eggs are greenish-white.

This bird is common on the Test, the Itchen, the Kennet, and wherever high reeds grow.

THE REED-SPARROW.

The REED-SPARROW (*Emberiza schœniclus*), or the Bunting of the Bulrushes; family, *Emberizidæ;* also known as the Reed Bunting, Black-headed Bunting, Black Bonnet, and Coaly Hood, is common enough on most of our rivers. The male bird, with his black head and throat and white collar, is a prominent object as he flits from reed to reed by the side of the river, lake, or marsh; he is also very fond of osier-beds, flags, and rushes. Not much of a songster; but a very persistent performer whilst his mate is sitting. Naumann gives his song, *Zja-til-tai, zississ-tai, zier-zipiss,* loud, and with a kind of stammering. Its call-note is like the word " *Tscheeh.*"

THE REED-SPARROW.

The bird is not very shy; indeed, we have seen the cock bird sitting on a reed uttering its call-note with a fly-fisher not fifteen yards off, and not at all scared by the line passing close to him. The nest is placed close to the ground, often under the grass which covers the banks of water-courses in the meadows. Eggs purplish pale brown, slotched and streaked. There are often two, or even three, broods in the year. The female, if disturbed when hatching, shuffles off as if her wing was broken, and tumbling about, endeavours to draw the intruder from the nest.

In the winter months it consorts with other buntings and pipits, and frequents the high lands near sheepfolds, and in very severe weather may be found about the farm-yards and rick-yards with the finches and sparrows. Its food is then chiefly graminivorous, whilst in summer it lives on insects, larvæ, and small snails.

THE BEARDED TITMOUSE.

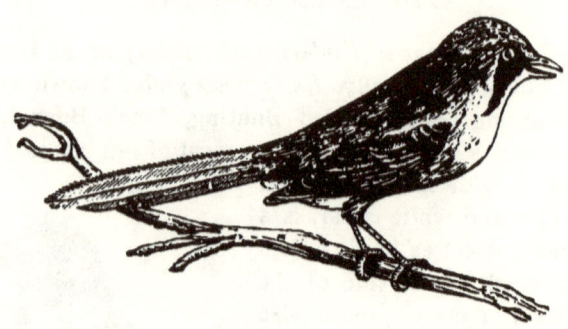

Another bird of the reeds but rarely seen is the beautiful BEARDED TITMOUSE (*Calamophilus biamarcus*); or, "The Friend of the Reeds." Order, *Passeres*; family, *Panuridæ*.

This bird is now chiefly confined to the Broads of Norfolk, more particularly to Hickling and Horsey Broads. Seebohm says that it is still found in some parts of Devonshire. It was formerly common enough on the banks of many of our slow-running rivers, and is, we believe, still occasionally met with on some parts of the Thames. Dresser calls it the Bearded Reedling, and in Norfolk it goes by the name of the Reed Pheasant. Bailey says: "The habits of this bird are quiet and sociable; they do not fear man much, and only when disturbed or menaced do they fly off to some distance or plunge into the reeds, uttering their call-note, *Thien-thien*." This cry somewhat resembles the silvery sound produced by twitching the strings of a mandoline.

If you should happen to come across one of these birds, put your binocular on it at once, and, if a male bird, look carefully at his beautiful plumage—the dark slate-coloured head with his black moustache, rather than beard, extending from the eye and ending in a point; the pinkish sides and beautiful fawn-coloured back and wings, with the lesser wing-covert grey; the long fawn-coloured tail. And watch him how he creeps up the stems of the reeds, sometimes head downwards, always on the alert—a lovely bird indeed.

THE MOOR-HEN.

What is that curious little black lump floating down the stream—looking like a bit of dubbing from a fly-fisher's book—black, red, and blue? But that it cannot be, as it utters a weak, cheepy cry. Put the binocular upon it, and see, it is a little rumpled feathery mass, with a red and blue head, looking utterly helpless and miserable. It is the young chick of the MOOR-HEN or WATER-HEN (*Gallinula*

THE MOOR-HEN.

chloropus—*i.e.*, the greenish-yellow-footed bird)—family, *Rallidæ*—which has just escaped from the nest, and somehow or other got into the stream, and is drifting away, too weak to resist the running water. The mother, you may be sure, is not far off. Hark! there is her cry. It sounds, however, far up the stream. She is a ventriloquist; for, look! out she comes from the flags opposite and close to her chick, with that peculiar jerk of her tail. If you remain quite still you will see she will get below the little helpless thing and guide it back amongst the sedges, where it will be out of harm's way.

Moor-hens guard their eggs and young with great care. Jesse says: "A crow was seen to alight on the edge of a piece of water near the nest of a moor-hen. Immediately the cock bird flew at him with such force as to knock him over, and continued the attack with such spirit that the crow soon took to his wings and flew off."

The moor-hen is found on almost every river, lake, or large sheet of water, and often on very small ponds, throughout the kingdom. It is one of the commonest of our water-birds. When disturbed it flies in a low, fluttering, scrambling manner, with its legs hanging down, and slides, as it were, with a long splash amongst the reeds, immediately commencing its peculiar call-note, *Crekre-kreck*. Seebohm says *Kek-ek-ek*, modulated at times into *Kik-ik-ik* or *Kerk-erk-erk*. Morris, in "The Earthly Paradise," says:—

> "And now the water-hen flies low,
> With feet that well-nigh touch the weeds.'

This bird, when frequenting ornamental waters near houses, becomes very tame, and some say is able to discriminate between friends and strangers, remaining quiescent and feeding when the usual inhabitants are about, but scuttling away at once on the appearance of a strange man or dog.

Like many other birds, the moor-hen, having once taken up a particular haunt, allows no other bird to interfere with him, fights bravely for his home, and will drive off all intruders. Although living as much on the water as the land, its feet are not webbed, and one sees it constantly seeking its food walking with slow and graceful steps over the meadows and lawns.

The bird is about 13 inches in length, and, what is unusual, the female is more vividly coloured than the male, and is also somewhat larger. The colour on the back is olive-green, with a red patch on the forehead; red beak with yellow tip; yellow-green legs, with a red ring above the joint. The upper part of the tail is dark olive, the under coverts white, which show when the bird jerks its tail as it comes out into the open water. Drayton, in his "Polyolbion," notices this :—

> " The coot bald, else clean black, that whitenesse it doth weare
> Upon the forehead starr'd, the water-hen doth wear
> Upon her little tayle, in one small feather set."

Mr. Howard Saunders ("Yarrell," fourth edition, vol. iii.) says that Dr. William Turner, writing in 1544, calls the bird a Water-hen, or a Mot-hen, or it should be moat-hen, as in the days of moated houses these birds much frequented the moats. The term moor-hen probably comes from moorish, a synonym for marshy. Spenser says, "The moorish Cote and soft sliding Breame" ("Faery Queene," B. iv.). The nest is a curious lump of reeds and rushes, sometimes on a mass of flags in the water, sometimes on a bough overhanging the water. Clare says :—

> "At distance from the water's edge,
> On hanging sallow's farthest stretch,
> The moor-hen 'gins her nest of sedge,
> Safe from destroying schoolboy's reach."

Some people, particularly river-keepers, assert that the moor-hen destroys the ova of trout. It may be so, to a certain extent; on the other hand, the bird compensates for this by destroying in great quantities, the larvæ of the dragon-fly and the water-beetles, both great devourers of the ova and fry of trout.

THE COOT.

The Coot, *Fulica atra* ("the dark sooty bird"), of the same family as the moor-hen, is well known to all frequenters of the water-side, as it is not only found in considerable numbers on all our still waters, lakes, and broads, but it frequents most of our slow-running chalk-streams and larger rivers—in fact, it abounds wherever there is water favourable to it. It is a conspicuous object, with its dark plumage and white forehead.

Colonel Hawker says: "All wild fowl seek the company of coots, because these birds are such good sentries to give the alarm by day, when the fowl generally sleep." Drayton, in the "Owl," says:—

"The brain-bald coot, a formal, witless ass."

Skelton, in his elegy to the death of Philip Sparrow, says:—

"And also the *mad* coote,
With a bald face to toote (pry)."

Hence the sayings, "As mad as a coot," "As bald as a coot."

THE COOT.

Why Skelton should call this bird mad, and Drayton a witless ass, is strange, as it is one of the most wary of birds, and is the first to give warning to all its feathered friends by its call-note, *Ko-ko*, oft repeated. Coots are splendid divers, and should there be any noise or strange appearance to which they are unaccustomed, they will instantly dive, and the surface of the water, a moment previous black with hundreds of these birds, will be like a glassy mirror :—

> "When the blue breast of the dipping coot
> Dives under, and all is mute."

The expectant observer will be astonished at their non-appearance; but they have all made their way under water to the nearest shelter, and after a while, when all is quiet, comes their call-note from the reeds, and one by one out they come, and the whole surface will again be teeming with black life.

That he is anything but a witless ass is proved by the mode in which this bird prevents the attack of birds of prey. In Dresser's "Birds of Europe," Lord Lilford communicates the following :—"It (the coot) is very common in winter on the lakes of Epirus, in which country I have several times observed the singular manner in which a flock of these birds defend themselves against the white-tailed eagle. On the appearance over them of one of these birds they collect in a dense body, and when the eagle stoops at them they throw up a sheet of water with their feet and completely baffle their enemy. In one instance, on a small lake near Butrinto, they so drenched the eagle that it was with difficulty he reached a tree on the shore not more than a hundred yards from the spot where he attacked them." These remarks corroborate what Sir Thomas Browne, of Norwich, when writing of British birds about 1635, says :—"Coots are in very great flocks on the broad waters. Upon the appearance of a kite or buzzard I have seen them unite from all parts of the shore in strange numbers, when, if the kite stoop near them, they will fling up and spread such a flash of water with their wings, that they will endanger the kite, and so keep him off again and again in open opposition."

Coots feed chiefly on aquatic insects, worms, slugs, land and water snails, and various water-plants—particularly the submerged leaves of the pond weeds (*Potamogeton*) and *Ranunculus aquaticus*. The river-keepers declare that this bird takes the small trout and the ova. It will no doubt eat small fish occasionally. Some years since, whilst watching the coots on the Lake of Lucerne, we saw one of these birds seize a bleak and eat it. It happened thus: There was a great scamper amongst a large shoal of bleak arising from the dash of a pike or trout, and one appeared to be injured, and kept jumping out of the water. A coot immediately rushed at it, seized it, and after killing it by hitting it on the surface of the water, swallowed it.

It was amusing to see some of the old male birds, too lazy to dive for their own dinner, carefully watching the younger ones busy at the bottom of the river, and the moment they rose to the surface with beaks full of weeds give chase, and, like the skua amongst the gulls, force them to relinquish their hard-earned meal, which these thieves then eat at their leisure. Half paddling, half flying, the coots run, as it were, on the surface of the water when chasing each other, bringing to mind the words of Burns:—

"The wanton coot the water skims;"

but when alarmed they can fly with great rapidity.

The coot is about eighteen inches long. The plumage is of a purply sooty-black, with a tinge of grey in certain lights; the secondary wing-feathers are tipped with white, forming a kind of bar. The beak is flesh-colour, the sides crimson, the patch on the forehead almost pure white, the legs, toes, and membrane green, an orange band above the tarsal-joint.

THE WATER-RAIL.

Another of the family *Rallidæ* is the WATER-RAIL (*Rallus Aquaticus*), Velvet-Runner, Shilty-Cock, Dar-Cock.

Being a very shy bird, it is not often seen, carefully concealing itself when danger is near. We have been fortunate enough to flush it occasionally both in Wiltshire and in Scotland. It flies much like the moor-hen, with hanging legs, to no great distance, and when alighting runs rapidly and seeks the shelter of high grass and reeds.

THE WATER-RAIL.

The beak is red, eyes hazel, top of the head, neck, back, and wing-coverts and upper surface of the body olive-brown, each feather having a dark centre; front of the neck and breast lead-grey, sides and flanks slaty barred with white, legs and toes brownish flesh-colour.

Seebohm says: "The water-rail is almost as exclusively a reed bird as the bittern or great reed-warbler. The one great object of its life appears to be to conceal itself. It threads its way through reed and sedge, only occasionally venturing to swim across a narrow piece of open water, and never exposing itself or venturing out to feed on the

THE WATER-RAIL.

grass in the neighbouring meadows until its movements are concealed by the shadows of evening."

Lord Lilford (*Field*, December 8, 1888) says that the water-rail has a considerable variety of notes—a shrill, twittering, long-drawn "skirl," to be heard at all times of the year, especially in the early morning or evening; also a sharp single note, somewhat resembling the twit of the pied woodpecker; also a low chuckle when in search of food, with a continual pecking of the tail, after the fashion of the common water-hen." He states also that a gun-shot or a distant roll of thunder will often set off the water-rails and other allied species screaming and twittering for minutes together. This is not an uncommon occurrence with other birds when alarmed by a loud noise.

THE WATER-OUZEL.

The WATER-OUZEL or DIPPER (*Cinclus Aquaticus*), one of the family *Cinclidæ*, is scarcely ever seen on our southern chalk-streams; but in Devonshire, Dorsetshire, Somersetshire, Herefordshire, and all through the the midland and northern counties, in Wales, and in Scotland the dipper

THE WATER-OUZEL.

is found on almost every brook or stream, flitting from stone to stone, now and again pouring forth his sweet, melodious song.

Stand still and watch him for a few moments, and you will see him disappear under the water:—

"In the osier-bank the ouzel, sitting,
Hath heard our steps, and away is flitting
From stone to stone as its glides along,
Then sinks in the stream with a broken song."

This disappearance is not by diving, but by gradual immersion. He is after his food, and will appear again in a few moments and fly to the nearest stone, or join his mate, who has made the neatest of moss nests under some hanging rock or archway of a bridge. Its movements and the mode of elevating its tail put one in mind of the wren; the white on the breast makes it a conspicuous object, and at once distinguishes it.

In some counties it goes by the provincial name of Bessie Ducker; in Cornwall it is the Water-Thrush; in Westmoreland the Water-Crow; in Scotland the Water-Piet; in Ireland it is known as the River-Pie. In some parts of England the common people believe that the dipper is the hen kingfisher. In the Highlands it is called the Water-Blacksmith.

The Duke of Argyll ("Unity of Nature," p. 81) says: "The dipper or water-ouzel is well known to ornithologists as one of the most curious and interesting of British birds. Its special habitat is clear mountain-streams. These it never leaves, except to visit the lakes into which or from which they flow;" and Seebohm says that "the haunts of the dipper are exclusively confined to the swift-flowing, rocky mountain-streams;" but this bird is not at all uncommon in some of the streams of the midland districts, viz., the Teme, the Corve, the Ony, and many others. Its haunts appear to us to be quick-flowing, rocky, and shallow streams, not necessarily mountain-streams. The Duke well describes it as moving about the bottom of the river as if it had no power to float, and floating on the top of the water as if it had no power to sink. St. John says: "In the coldest days of winter I have seen him alight in a great pool, and with outstretched wings recline for a few moments on the water, uttering a most sweet and merry song; then rising in the air he wheels round and round for a minute or two, repeating his song as he flies back to some accustomed stone."

There has been a question as to whether the dipper, or, as he is called in Scotland, the water-crow or kingfisher, is a fish-eater? Macgillivray, who examined their stomachs, found only beetles, fresh-water shells, caddisworms, larvæ of the *Phryganidæ, Libellulæ*, (dragon-flies), and water-beetles. All these insects in their larva state are very destructive to fish-spawn, particularly trout-ova.

In Scotland the bird is persecuted, in the belief that it destroys great quantities of salmon and trout ova, as well as the fry. Sir W. Jardine, however, denies this, and says that the spawn of either salmon or trout have never been detected in their stomachs.

Professor Metzzer has on different occasions found the bones of fish, chiefly minnows and bullheads, in the stomachs of the old birds. Herr Müller says: "I not only learned that the water-ouzel fishes, but also that in summer, spite of his great liking for water-insects, which abound in the Schwalm, he evinces partiality for fish diet."

There can be no doubt that the chief food of this bird consists of the caddis-worm and larvæ of beetles, dragonflies, and the like, all great feeders on the spawn and fry, and although it does occasionally take small fish, yet there is the usual compensation which always exists in such cases.

The following interesting account taken from the *Field*, and written by Mr. Bartlett, superintendent of the Zoological Gardens, is worthy of attention :—

"Year after year I have tried without success to rear from the nest these very interesting and singular birds, and notwithstanding repeated failures, I have not only continued in this endeavour, but have induced others to make the attempt. In these efforts I have been aided by several, and among others, Mr. R. J. L. Price, of Merionethshire, a Fellow of the Society. This gentleman kindly forwarded the nests of young birds, and from time to time, by trying almost every kind of insect and other food, I succeeded for a while to rear the birds, but just when our efforts appeared likely to succeed a change would take place, and the birds would die one after another. Sometimes they would get too wet and die apparently of cramp ; others that had been kept away from the water wasted and died of exhaustion. It was quite evident that we had not discovered a food that suited them ; they had been tried with the usual food for most insect-eating birds, such as scraped beef and hard-boiled eggs, ant-eggs, meal-worms, spiders, flies, beetles, aquatic snails, shrimps, salmon-spawn, and many other mixtures, but all failed until the

20th of May, when my clerk and assistant, Mr. Arthur Thomson, who had taken as much interest in rearing these birds as myself, hit upon the idea of scalding the meal-worms, and tried it. It was soon apparent that in this condition the meal-worms could be digested, while in a raw or living state they (especially their hard skins) would pass through the birds in a hard and undigested condition. From this moment we had but little trouble. The birds fed greedily upon the half-boiled meal-worms, and we soon found them ready to leave the nest, and accordingly fitted up a cage, having the nest under a rock in one corner and a shallow pan at the other end of the cage, in which the birds soon began to dive and swim about. They are now about six or seven weeks old, feed themselves, or nearly so, being excessively tame, and they still come to be fed by hand. Since they have taken to feed themselves the food has been greatly varied by introducing caddis-worms and other aquatic insects of small size found among the weeds; this affords them much amusement, and they throw up castings, or pellets, after the manner of raptorial birds. The pellets consist of the parts of the insects that are not digested. It is most interesting to watch their movements, bobbing up and down, flying from place to place, and diving under water and extracting the caddis from its curious covering. I can no longer doubt the charges brought from time to time against our pets of appropriating a small portion of the young trout or salmon, for they are most expert fishers; but I feel perfectly satisfied they do not eat the roe or spawn of fish. As I have before stated, unless there is some movement, these birds do not eat anything they find. In diving, the dipper uses its wings as though it was flying under water, and has to exert considerable force to remain under long enough to capture its food; it is so buoyant that it floats to the surface like a cork. The song of the water-ouzel is said to be louder, but in other respects much resembles the wren. Our young birds already give indications of their vocal music. I can find no very correct description of the movements of the dipper; I take, therefore, this opportunity of stating that the bird runs about rapidly after the fashion of a starling. It jumps or hops a considerable distance; it flies well, and swims like a duck. I have six of them altogether. The birds are from two nests—one contained three, the other four birds."

This question of eating fish is, without doubt, an important one, and however much we admire the bird, the weight of evidence is against it.

Mr. A. D. Bartlett, in the *Field*, says:—" However unwilling I may be to render some of my pets to be regarded as the enemies of fishermen, the truth must be told. In May 1869 I obtained my first living water-ouzel. Since that time I have had upwards of twenty of these birds. Some of them I have reared from the nest, and I fed them upon boiled meal-worms, the larvæ of the caddis-fly, and other insect food; but as soon as they were able to feed themselves and took to the water, they caught and fed upon very small fish, especially young minnows. I found them rather expensive pets, having to provide for a family of four, as they caught and devoured several dozen daily, and seemed to prefer live fish to all other food. I am not pleased to confess this, and I hope it may not cause the birds to be unmercifully killed, as I feel sure that these birds are useful, feeding as they do upon insects as soon as the young fish are too large for their tiny throats."

As to whether they take spawn, a correspondent signing himself "Nahanite" writes:—

"Sir,—Some thirty years or more this subject was very carefully considered by the members of the now defunct Dublin Natural History Society. The birds were carefully watched when at work on the spawning-beds, and afterwards shot and dissected, and as well as I can now recollect in no case was spawn found, or if there was, very little; while numerous caddis-worms and other insects destructive to spawn were found to be what they had been collecting, these depredators frequenting the beds to destroy the spawn, and the ouzels going there to devour them."

The nest is somewhat like a wren's nest, only a little larger, and chiefly made of moss growing in the locality, usually placed under a shelving rock or under the arch of a bridge. We found one attached to a half-brick under the arch of a conduit conveying pure water to the town below. It is also fond of the side of the water-wheel of a mill. Mr. Thomas Edward, the Banff naturalist, sent the following to the *Zoologist* :—

"DIPPER'S NEST ON THE TOP OF A BOULDER.—That many birds build their nests in what seem to us odd places is well known, and perhaps in this respect the dipper is one of the most eccentric. Not to mention cases of which I have heard and read, I have myself seen their nests in very extraordinary—nay, almost incredible—places. In this I think the one I have now to notice will in a measure bear me out. In a river near here, and about midway in the stream, there lies, amongst others, a small boulder, which shows a foot or so above the usual current, and has a very slight depression on the top. Here a pair of water-crows commenced their nest, and by some extraordinary means succeeded in rearing a home for themselves and their young. When finished, although it heightened the appearance of the boulder considerably, yet, from the shape and colour of the materials used, the nest could scarcely be distinguished from the boulder. It was discovered by the merest chance by a gentleman whilst fishing. The river at this spot is very bare, having no bank of any height, bush, nor tree near; yet the nest, although completely exposed to every gust of wind and rain, nevertheless withstood the fury of the elements for at least three months. By this time a family of four had been reared, and with their parents had departed elsewhere. The nest would have stood longer, but was removed to be preserved as a memorial of the strange place where built. The removal, however, proved a much more difficult matter than was anticipated. On the attempt being made the nest was found to be so firmly attached, cemented on, as it were, to the stone, that it took some considerable time and trouble to detach it. On being minutely examined it appeared from its construction to be impervious to rain."

Mr. G. Rooper records a dipper's nest in a tree, and Mr.

Whitaker, one made in the middle of a tangle of weeds which had attached itself to a low branch of a tree overhanging the stream.

As we have remarked above, this bird is at once recognised from all other birds of the rivers and reeds by his white breast; the back is brownish-black, the head and neck umber-brown, the under parts chestnut.

THE DABCHICK.

The LITTLE GREBE or DABCHICK (*Podiceps minor*)—family, *Podicipedidæ*—has various provincial names, as Dabchick, Small Doucher, Dabber, Dive-Dapper.

THE LITTLE GREBE—DABCHICK.

This bird, like the moor-hen, is found on all still waters, large or small, and on most, if not all, of our rivers, particularly on our chalk-streams. It is most destructive to and devours vast quantities of fry and yearling fish of all kinds, as well as the ova. The following from the *Fishing Gazette* shows their propensity for this kind of food:—

"On many trout-streams dabchicks may now be seen disporting themselves in places where they are not seen at any other period of the year. By a curious coincidence, these same places are just where the trout are spawning. I recently counted no less than six dabchicks on a lonely shallow in the Kennet, feeding bravely on the bottom. Each bird dived about ten times in a minute, and if each dive resulted in the destruction of a trout-egg, it will, by a simple arithmetical calculation, be discovered that trout-eggs were being destroyed at the rate of 3600 per hour.

THE DABCHICK.

"I recently shot one of these birds (not one of the six), and examined its crop. Therein I found three trout-ova, a miller's thumb, a large number of water-shrimps, and various other odds and ends not easily classified. I have not the least doubt that later in the year, when the trout-fry are about, many thousands of them go down the throats of these pretty little pests. When not eating fry or ova, they fill themselves with food which, from our point of view, is the rightful property of *trout*," viz., the larvæ of the May-fly. Sometimes, however, they take other fish, and occasionally catch a Tartar. A dead dabchick has been found choked by a large bull-head.

A pair of dabchicks will do much more damage to a river than a pair of otters. The amount of small fish they take is perfectly astounding. A pair of dabchicks confined in the Fish-House of the Zoological Gardens cost the Society a considerable sum per week in providing small fish for them. Think of this, ye breeders of trout, and do not be surprised at the absence of fish in your streams if dabchicks are plentiful. Fishing a well-preserved river the other day, we counted no less than seven pairs of dabchicks feeding as hard as they could fish in the space of a mile of the river. Much as we love to see all the birds of the stream enjoying themselves, we are afraid this little rascal is too much even for us.

It is a bold, fearless bird, but very wary. Whilst fishing on the Itchen this last summer, the keeper suddenly put the net into the water and fished up a dabchick. He was a male, in full breeding plumage, and beautiful he looked; but the moment he was caught he went at our fingers with his sharp-pointed beak, not appearing at all scared.

This bird is often seen on small ponds, but it prefers open water. We once saw one on the Serpentine, and the *Zoologist* records a nest on the Round Pond in Kensington Gardens some years back. In full breeding plumage the male has a deep reddish-chestnut neck, with the breast a greyish-white; upper surface of the body very dark brown, dark-green legs, eyes a reddish-brown. The bird builds rather a large nest of flags and reeds, and often places it on a mass of cut stagnant weeds, and also among the reeds and half-immersed herbage at the sides of the ponds and streams. The eggs are generally from five to seven in number, whitish; and the birds, male and female, take turns in the process of incubation, and when absent the eggs are carefully covered over with grass and flags, which they do with their beaks. Their short wings make them bad flyers. Pope gives this description of their mode of progression:—

> "As when a dabchick waddles through the copse,
> On feet and wings, and flies, and wades, and hops."

But they are wonderful divers, and on the slightest alarm are down below the surface in a moment. Drayton, in his "Polyolbion," Song xxv., says:—

> "The diving dabchick here amongst the rest you see,
> Now up, now down again, that hard it is to prove,
> Whether under water most it liveth, or above."

The dabchick has its place in Mythology. In Ovid's "Metamorphoses" it is stated that Œsacus was transformed into a dive-dapper. That is to say, the Latin word *mergus* is so translated; but in Lempriere *mergus* is a cormorant. The word *mergus* really means a diver; and Ovid could not give a better example of a diver than the dabchick.

THE MALLARD.

All our large southern streams run through rich valleys and water meadows well adapted for the nesting of our water-birds, the wild duck amongst them. Every one knows the WILD DUCK (MALLARD) (*Anas Boschas*); of the family *Anatidæ*.

In sporting language the mallard is the male of the wild duck; but in the language of natural history the mallard is now the specific name of both male and female to distinguish it from other forms of wild ducks. Notwithstanding this, we, as sportsmen, must continue on the old tracks and call the male bird the mallard and the female the duck. Seebohm says: "*Mallard* is a French word meaning drake, in contradistinction to *canard*, which means a duck." Probably the word mallard is a corruption of male canard. The wild duck is supposed to be the origin of our domesticated ducks, but it is a circumstance worth remarking that the wild bird pairs, while the domestic bird is polygamous.

It is in the spring-tide when its plumage is in perfection. Just throw your binocular upon him as he stands

THE MALLARD.

erect in the water-meadows watching your movements some hundred yards or so from the river-side, half inclined to take wing, yet loth to leave his wife nestling close at hand. Look at the splendid emerald-green on his neck flashing in the sun's rays, the varied plumage of the back and breast, chestnut and grey:—

> "The mallard, young and gay,
> Whose green and azure brighten in the sun."

How beautiful he is! And where is his dusky mate?

Hard sitting on her eggs in some reeds close by, or under the shadow of some overhanging tree, or may be in the hollow of some pollard willow, or even at times in some cast-off nest of some other bird higher up amongst the branches. When thus placed we may wonder how the old bird conveys her young ones to the water. Some say she carries them down on her back; but much more likely in the same manner as the woodcock—between her thighs —or the young may attach themselves under the pinion, something in the same manner as the dabchick.

Jesse ("Gleanings," p. 181) says that "at a place in North Wales some wild ducks had their nests on trees. The birds had been frequently watched whilst conveying their young to the ground, and in every instance one of the wings of the duck appeared to be closed, whilst she flapped rapidly with the other, evidently for the purpose of breaking her fall. She always alighted near the foot of the tree, thus descending nearly perpendicularly."

However that may be, the first duty of the mother is to convey her brood to the reeds by the river-side, or to some secluded piece of water not likely to be disturbed, and there hide them carefully all day. Should anything, however, occur to frighten or cause alarm, what a fuss she will make, flapping along the water as if her wing were broken, or hobbling along the grass as if desperately hurt, and so allure her enemy, whatever it may be, from her brood! Last year we happened to wade into a mass of flags by the river-side, when out burst a duck, almost touching us, and down she went, as if wounded, into the stream. We remained very quiet for two or three minutes, and then came *Cheep, cheep, cheep* at our feet. On looking down, there were half a dozen little brownish-black mites amongst the reeds. The moment the old mother heard the voice of her children her antics began. First she commenced flapping and gliding all along the water; then she got on to the grass and flew a little way, and tumbled over as if with a broken leg. Then she got up and flew round, and again tumbled into the water, all the time quacking loudly. When this did not move us, she rose and flew about a hundred yards down-stream, and then began beating the water with her wings, diving and flopping about in all directions. This, we thought, will never do; we got away from the young and followed her down. So delighted was she to find her ruse succeeding, that on she went, flap, flap, flap; so the only thing to do was to get rid of her by walking clean away from the bank, and when some two or three hundred yards from her up she got, made a circle round, saw the coast clear, and down she went into the meadow, and then crept along close to the ground till she got to her brood, there to remain till eventide, and then to lead her little ones to revel on some undisturbed reach of the river, where

"The dark trout spreads his waning, O !"

A very curious change occurs about the end of May in the drake. Beautiful as his plumage was in the beginning and middle of the month, it suddenly begins to alter, commencing on the breast and back; in a few days the curly feathers of the tail fall out, and by the end of June all the lovely green plumage is mottled with grey, and by the 6th of July he has put on almost completely the plumage of the female. This continues till about the middle of August, when another change takes place, and by the middle of October he reappears in all his pride of beauty.

In the *Zoologist* of November 1888 it is stated that young wild ducks are very fond of the larvæ of the *Phryganidæ*, large quantities having been found in their stomachs.

THE SWAN.

The SWAN (*Cygnus olor;* family, *Anatidæ*) is a very familiar object to the frequenters of the Thames-side. Tuberville, who wrote his sonnet to the Thames in the early part of the seventeenth century, says :—

"Thou stateley streame that with the swelling tide
  'Gainst London walls incessantly dothe beate,
Thou Temes (I say) where barge and bote doth ride,
And snow-white swans do fish for neadful meate."

Some declare that a great deal of this "neadful meate" consists of the spawn of fish, but in this respect the bird is probably much vilified. It certainly clears away a number of noxious water-plants which would in a measure choke up our streams, particularly the *Elodia canadensis*, or water-thyme, a North American plant, which bade fair at one time to become a national nuisance.

A male swan in full plumage is a magnificent object. Milton in "Paradise Lost" describes how—

> "The swan with arched neck,
> Between her white wings mantling, proudly rows
> Her state with oary feet."

And the following is Thomson's description of the male bird during the nesting season:—

> "The stately sailing swan
> Gives out his snowy plumage to the gale,
> And, arching proud his neck, with oary feet
> Bears forward fierce, and guards his osier isle,
> Protective of his young."

Yarrell truly remarks "that the swan is perhaps, of all others, the most beautiful living ornament of our rivers and lakes."

The swan so familiar to us is the mute swan (*Cygnus olor*), a different species to the wild swan, hooper, elk, or whistling swan (*Cygnus musicus*), and easily recognised by the beak. The mute swan has a *rich red orange beak with a black base* and a large black knob. The wild swan, or hooper, has a *black beak with a bright yellow base*, with little or no knob (see Figs. 1 and 2).

Swans are now rarely met with below Chelsea Bridge; but in former days they were often seen in the Pool below London Bridge.

Our mute swan is a native of Cyprus, and was introduced into this country some time in the twelfth century.

The swan has a mythological history. It is the bird of Apollo. Cycnus, the son of Hyrie and Apollo, having thrown himself off Mount Teumesus in a fit of resentment, was transformed into a swan. Ovid tells of another Cycnus, son of Sthenelus, and closely related to Phaeton, who was, after Phaeton's death, transformed into a swan:—

> "His hair transformed to down, his fingers meet
> In shining films, and shape his oary feet.
> From both his sides the wings and feathers break,
> And from his mouth proceeds a blunted beak.
> All Cycnus now unto a swan was turned,
> Who, still remembering how his kinsman burned,
> In solitary pools and lakes retires,
> And loves the waters as opposed to fires."

In poetry the tradition that the swan sings as he dies

FIG. 1.—HEAD OF THE MUTE SWAN.   FIG. 2.—HEAD OF THE WILD SWAN.

continues to the present time. Spenser, in his "Ruins of Time: Fall of Rome," says:—

> "There he most sweetly sung his prophecie
> Of his owne death in doleful elegie."

Sir Christopher Hatton says:—

> "The silver swan, who living has no note,
> When death approach'd unlocked her silent throat;
> Leaning her breast against the reedy shore,
> Thus sang her first and last and sang no more:
> Farewell all joys! O Death, come close my eyes!
> More geese than swans now live, more fools than wise!"

This was set to music as a madrigal by Orlando Gibbons in 1612.

And Mrs. Hemans :—

> "'What is that, mother?' 'The swan, my love;
> He is floating down from his native grove;
> Death darkens his eye and unplumes his wings,
> Yet his sweetest song is the last he sings.'"

Byron also says in "The Isles of Greece:"—

> "There, swan-like, let me sing and die."

And Tennyson, in his "Ode to the Dying Swan," says :—

> "The wild swan's death-hymn took the soul
> Of that waste place with joy."

We do not suppose for a moment that any one of these poets ever heard this death-song. It is only perpetuating a myth which had its origin in the supposition that the soul of Orpheus was transmigrated into a swan. The following lines give a much more beautiful application :—

> "The silver swans, no more than other fowl,
>   With tuneful notes, presage impending death;
>   The notion of their dying tuneful breath
> Was meant as emblem of a pious soul;
> Such, whose fair life, white as their snowy down,
>   Not stain'd with the opprobrious marks of vice,
>   Arriving at the gates of Paradise,
> Their end, with joyful resignation crown."

The swan has no song, but the wild swan has a loud kind of trumpeting voice, which, when oft repeated by a large flock, has somewhat of a musical note, and has been compared to a pack of hounds in full cry.

In the fourth edition of Yarrell's "British Birds," edited by Mr. Howard Saunders, there is an interesting account of the mute swan, and the swans' marks of the various proprietors and corporations who possess these birds. The swans on the Thames, *i.e.*, to a certain distance, belong chiefly to the Queen, the Vintners', and the Dyers' Companies.

The corporations of different towns have also privileges in connection with swans—Norwich, for instance, where they collect a certain number and fatten them for the table,

THE SWAN.

and when sent to the different receivers, are accompanied with the following instructions for cooking, written by the Rev. J. C. Matchell:—

To Roast a Swan.

Take three pounds of beef, beat fine in a mortar,
Put it into the swan—that is, when you have caught her;
Some pepper, salt, mace, some nutmeg, an onion
Will heighten the flavour in gourmand's opinion.
Then tie it up tight with a small piece of tape,
That the gravy and other things may not escape;
A meat paste, rather stiff, should be laid on the breast,
And some whited-brown paper should cover the rest;
Fifteen minutes, at least, ere the swan you take down,
Pull the paste off the bird, that the breast may get brown.

The Gravy.

To a gravy of beef, good and strong, I opine
You'll be right if you add half a pint of port wine;
Pour this through the swan—yes, quite through the belly,
Then serve the whole up with some hot currant jelly.

N.B.—The swan must not be skinned.

THE RED-BREASTED MERGANSER.

The RED-BREASTED MERGANSER (*Mergus serrator*), or, as he is called in Scotland, the Saw-Bill, Saw-Neb, or Diving-Goose, frequents many of the lochs of Scotland and Ireland during the breeding season, and when the young are hatched the mother leads them, after a time, to the rivers running from these lakes. On their way to the sea-coast and in the pools the old one and the brood may be constantly observed diving and fishing.

HEAD OF THE MERGANSER.

To the salmon-angler this bird is a *bête noir*. The local fishermen will tell you that wherever you see the mergansers feeding, no salmon will rise, and we ourselves have had practical experience of this.

Mr. Arch. Harper, Brawl Castle, Thurso, N.B., writes:—
"I shot several mergansers within the last fortnight, and being anxious to know what they feed on at all seasons of the year, I opened the stomachs of them all. They, without exception, had the same quality, although the quantity of food varied, viz., from three to seven pars in each. Some of the pars were six inches long. I have now shot and examined these birds at all seasons of the year on this river, and they invariably have par in their stomachs, and nothing else."

Still, destructive to sport and great gobbler of all kinds of fry as he is, the male bird is as handsome as he can be. Just put your glass upon him when you happen to catch him quietly resting on a pool and not alarmed; see what a splendid plumage, his dark head and back, resplendent with green and purple hue, the beautiful crest and scarlet serrated beak, the white throat, and speckled breast of buff and black, mottled black and white wing-coverts, orange legs and feet; notice particularly his wicked red eye, vicious to a degree, as if he would dispute with you the possession of the pool, though wary enough to get out of your way. The female is brown, with whitish-grey breast. The male bird never accompanies the female and brood in their peregrinations. It appears as soon as the young are hatched he takes himself off to the sea-coast, leaving his mate to do all the family duties.

Another bird of this family, the GOOSANDER (*Mergus merganser*), is occasionally found breeding on some of the lochs in Scotland; but it is a very rare bird, and, happily, does not interfere with the sport of the rod-fisher.

THE TEAL.

The TEAL (*Anas crecca*) is one of the smallest of our ducks, and is very prettily marked. It is not often seen on our rivers of the south, but frequents most of the

THE TEAL.

large pieces of water in different parts of the country where it breeds. On the broads of Norfolk it is more abundant. In Scotland one often comes on a brood when fishing the mountain tarns and streams. It is first rate for the table. The male bird has a very beautiful plumage. The top of the head is a rich chestnut-brown with buffish stripes, with a broad patch of rich green extending backwards; cheeks and side of the neck rich chestnut; back of the neck and back a mixture of blackish-grey and white. All the smaller wing-coverts ash-brown, the larger tipped with white, and the secondaries a mixture of velvet-black, green, and purple tipped with white; the front of the neck chestnut, lower part covered with dark spots, with a tinge of purple; legs and toes brown-grey.

Bewick says: "This beautiful little duck seldom exceeds 11 ounces in weight, or measures more than $14\frac{1}{2}$ inches in length."

THE TEAL.

CHAPTER III.

THE LAPWING.

THE LAPWING or PEEWIT (*Vanellus Cristatus*), one of the family *Charadriidæ* (charadrius, according to Pliny, being a bird the seeing of which cures one of the jaundice), migrates to this country early in spring, and many of them take up their abode in the marshes and water-meadows which border many of our rivers, not only for breeding purposes, but also for the abundant supply of food. It is a very wary bird, and our attention is soon drawn to it as it approaches us by its familiar cry, *Peewit, peewit.*

No bird is more crafty in luring the intruder from the nest. The female slips off quietly, and runs for some time close to the ground, bringing to mind Shakespeare's lines in " Much Ado About Nothing : "—

> " Look where Beatrice, like a lapwing, runs
> Close by the ground to hear our conference ; "

and then both the male and female come screaming round with their "sounding flight and wailing cry," wheeling upwards, and then passing so close that the rustle of the wings is plainly audible ; but as the nest is approached the birds become quite silent :—

> " But if where all the dappled treasure lies
> He bends his steps, no more she round him flies ;
> Forlorn, despairing of a mother's skill,
> Silent and sad she seeks the distant hill."

Should she be disturbed when with her brood, what mishaps she feigns, to draw off the attention of the intruder, be it dog or man, flapping along the ground with a broken wing or fractured leg, falling flat as if dead, and many other strange antics !

It has many provincial names, Lymptwigg in Devonshire, Peweep in Norfolk, Puit in Essex, Tuet in Lancashire, Flopwing in Beds. In Ireland it is known as Phillipene. In Scotland its general name is Peaseweep.

> " Peaseweep, peaseweep,
> Harry my nest and gar me greet,"

is an old Scotch nursery rhyme. In that country it has a bad name, as it was supposed by its cry and movements to have guided the troopers of Claverhouse to the hiding-places of the Covenanters :—

> " But, though the pitying sun withdraws his light,
> The lapwing's clamorous whoop attends their flight,
> Pursues their steps where'er the wanderers go,
> Till the shrill scream betrays them to their foe."

On the other hand, Sir Hercules Tyrrwhit, having been severely wounded, was saved by his followers being directed to the spot where he lay by the cries of these birds and their hovering over him, and in gratitude the family bear three peewits for their arms.

Chaucer calls it

> "The false lapwing, full of trickerie."

And Shakespeare, in "Measure for Measure:"—

> "Though it is my familiar sin
> With maids to seem the lapwing, and to jest
> Tongue far from heart."

The peewit has an ancient history. Ovid tells us that Tereus, king of Thrace, for his cruel behaviour to Philomela, sister of his wife Procne, was made unwittingly to devour his own offspring, and was then transformed into a lapwing, to be for ever wailing and restless, searching for his lost child.

The following legend is taken from the Danish ("Notes and Queries," vol. x. p. 49):—"While our Lord hung yet upon the cross there came three birds flying over. The first was the stork, who cried, 'Styrk ham! styrk ham!' (strengthen Him), and hence the bird's name and the blessings which go with her. The second cried, 'Sval ham! sval ham!' (cool or refresh Him), so she came to be called the swallow, and is also a bird of blessing. But the last was the weep (peewit), who shrieked, 'Pun ham! pun ham!' (pine Him, make Him suffer), and therefore she is accursed for ever down to the last day."

The lapwing is a handsome bird, with his black head and crest, and green, glossy back, with white under-parts; and as he turns and twists about in his varied flights, these white feathers show very prominently:—

> "The white wing plover wheels
> Her sounding flight."

As golden plover is considered a great delicacy for the table, the lapwing is often sold for it. They may be easily distinguished from each other by the feet. The lapwing has a hind toe, while the golden plover is without one.

THE SNIPE.

THE SNIPE.

The SNIPE (*Scolopax Gallinago;* family, *Scolopacidæ*). In the early spring, as one strolls by the side of the stream, a curious sort of drumming sound is often heard in the air above, which to the uninitiated is somewhat puzzling; but stand still for a few moments and you will see a bird (the male snipe) suddenly mount from the meadow to a considerable height in sweeps, then suddenly stop, open the wings and tail like a fan, and then fall in short zigzag flights, producing this peculiar humming or drumming, and when the bird arrives about twenty yards from the ground the tail closes, the drumming ceases, and on touching the earth, he makes a short run to where his mate is lying couchant. This peculiarity has given the snipe the name of Heather-Bleater in Scotland, Moor-Lamb in Lincolnshire, Air-Goat in Wales, *Chévre Volante* in France, *Himmel Zeige* in Germany.

The cause of this bleating or drumming sound has been the subject of considerable controversy. It was supposed for a long time to be produced by the wings in its downward flight; but Mr. W. Meyer, of Stockholm, made a series of experiments to prove it was due to the stiff vibration of the outer tail-feathers. Mr. Hancock, however, doubts this, and considers it the result of the strong action of the wings. From our own observations, it is probable that both wings and tail assist. It is noticeable that when the tail is closed on nearing the ground in the descending flight the drumming almost ceases. Seebohm, however, is inclined to the opinion that it is produced by the vocal organs, and is analogous to the trill of the stints and other sandpipers.

In Scotland the shepherds say that when the snipe drums the weather will be fine. If the opportunity occurs, look through the binocular and notice what an elegant form this bird has, how its long bill and green-brown legs agree with the beautiful brown, buff, and white plumage of the body; note its lustrous eye, and, if resting quite close to the earth, how it sits with its bill straight up and the body couched in the grass.

Drayton records how, in his time, this bird was considered a great delicacy, and calls it the "pallet-pleasing snite," and in France it was considered not only a delicacy, but a means of judging of the excellency of wine—

> " Le bécasseau est de fort bon manger,
> Du quel la chair resueille l'appetet,
> Il est oyseau, passager et petit,
> Et par son goust fait des vins bien juger."

THE HERON.

The HERON (*Ardœa Cinerea;* family, *Ardeidœ*), as he wends his way at eventide to take his supper on some shallow on the river, soon lets you know he has discovered your presence by his harsh alarm-note, *Krank, krank!* He flies apparently slowly with a long flapping wing, legs stretched out behind, and neck bent back, but he goes along much more rapidly than one supposes. Now and again we come upon him on the river-side in the daytime, and if approached cautiously without disturbing him —for he is very wary—he is worth looking at through your binocular. Notice how still he stands, his quick yellow eye always on the watch :—

THE HERON.

> "Lo! there the hermit of the water,
>   The ghost of ages dim,
> The fisher of the solitudes,
>   Stands by the river's brim."

His white forehead, his spotted neck, and long feathers at the back of the head; his pointed beak, so well adapted for its purpose; his blue, ash-grey plumage, with white under-parts and short tail; his long, greenish-yellow legs, enabling him to wade pretty deep in the water; and mark how cautiously he steps, scarcely moving the water as he lifts one leg in advance.

Compare his movements on the sea-shore when the tide

THE HERON.

is out with those of his associates around him, the curlews, the oyster-catchers, gulls, and stints; how busy they all are running about in every direction searching for their food, whilst this "solitary sentinel of the shore" stands erect, not deigning to notice them, but quietly waiting for some wandering fish or eel to come within reach of the unerring dart of his powerful beak.

No doubt in the breeding season the heron takes a goodly number of small fry, and even fish of a considerable size; but he is very partial to frogs and other inhabitants of the water—even will not despise the young coot or water-hen, or even the young of the water-vole. He has also a particular relish for eels. Drayton says :—

> "The herns
> Can fetch with their long necks, out of rush and reed,
> Snigs fry and yellow frogs, whereon they often feed."

And Burns, in his " Elegy on Captain Henderson : "—

> "Ye fisher herons watching eels."

And Leyden :—

> " Long-necked heron, dread of nimble eels "

Pontoppidan states that the heron likes small eels in preference to large, for this reason : they can be swallowed, give nourishment, pass through the intestines all uninjured, and then be swallowed again, and so on *ad infinitum.* This idea was prevalent in Ireland for a long time. However, herons, in their gluttony, sometimes endeavour to swallow large eels, much to their cost, as recorded in the fourth edition of Yarrell, where there is a drawing of a heron having pierced a large eel, which has twisted round the bird's neck, and so strangled it.

The heron is inclined to strike at any moving object passing within range. A few years ago we examined a fine salmon of twenty pounds found dead in the river, death having evidently been caused by the powerful beak of this bird having penetrated the brain.

The heron was formerly in great request, and very strictly preserved, not only for the purpose of sport, but for the requirements of the table. It was one of the chief pursuits in falconry, and gave more exciting flights than any other bird :—

> "The hern
> Upon the bank of some small purling brook
> Observant stands, to take his scaly prize,
> Himself another's game."

As an article of food it was much prized, and held a place in all great feasts. Walter Scott, in " The Lay of the Last Minstrel," giving an account of Lady Margaret's bridal feast, says :—

> "Pages, with ready blade, were there
> The mighty meat to carve and share,
> O'er capon, heron, stew, and crane,
> And princely peacock's gilded train."

At the installation of one of the Archbishops of York no less than four hundred heronshaws formed part of the feast.

Harting, in "Ornithology of Shakespeare," states that one of the last records of its appearance at table was at a feast given by the executors of Thomas Sutton, the founder of the London Charter House, 18th May 1812, in the hall of the Stationers' Company. For this repast were provided thirty-two neats' tongues, forty stone of beef, twenty-four marrow-bones, one lamb, forty-six capons, thirty-two geese, four pheasants, twelve pheasant pullets, twelve godwits, twenty-four rabbits, six *hearnshaws*, &c.

There are many heronries still existing in different parts of the United Kingdom, and there is an old tradition that the first chick hatched is always dropped from the nest :—

> "The heron from the ash's top
> The eldest of its young lets drop,
> As if it, stork-like, did pretend
> That tribute to its Lord to send."
> —ANDREW MARVEL, "Ode to Fairfax."

THE BITTERN.

The BITTERN (*Botaurus stellaris;* family, *Ardeidæ*) has many provincial names; Butter-Bump, Bog-Bumper, Bull o' the Bog, Mire-Drum, &c.

This bird, now so rarely seen, was, from its solitary, shy habits and its unearthly voice, formerly the terror of the inmates of the outlying farms on the fens and of benighted rustics; its booming being more like the bellowing of an enraged bull, and when heard, presaged death to some one of the family or misfortune to their herds.

Goldsmith ("Animated Nature") says: "I remember, in the place where I was a boy, with what terror the bird's note affected the whole village; they considered it the presage of some sad event, and generally found one, or made one to succeed it. . . . If any person in the neighbourhood died, they supposed it could not be otherwise for the night-raven had foretold it." (Willoughby considers the bittern and the night-raven the same bird). Another writes: "Those who have walked on an evening by the sedgy sides of unfrequented rivers must remember a variety of notes from the different water-fowl—the loud scream of the wild goose, the croaking of the mallard, the whining

HEAD OF BITTERN.

of the lapwing, and the tremulous neighing of the jack-snipe; but of all those sounds there is none so dismally hollow as the booming of the bittern. It is impossible for words to give to those who have not heard this evening-call an adequate idea of its solemnity; it is like the interrupted bellowing of a bull, but hollower and louder, and is heard at a mile distance as if issuing from some formidable being that resided at the bottom of the water."

This booming was formerly supposed to be produced by the bird putting his bill into the mud and then blowing through it; hence the provincial name, "Mire-Drum."

Chaucer writes ("Wife of Bath"):—

"And as a Bittore bumbleth in the mud."

Drayton says :—

> "The buzzin bittern sits, which through his hollow bill
> A sudden bellowing sends, which many times doth fill
> The neighbouring marsh with noise as though a bull did roare."

Others have supposed the noise was made by the bird thrusting his bill into a reed and blowing through it. Thus Dryden :—

> "Then to the water's brink she laid her head,
> And as a bittern bumps within a reed."

The bird has a very beautiful plumage—brownish-buff, crossed with brownish-black lines, head and nape of neck darker brown, beak greenish-yellow, eye yellow-green, legs and feet yellow-green.

THE COMMON SANDPIPER.

In strolling along the banks of a river or shores of a lake a small brownish bird will often be perceived running along the sandy margin, or starting in short, jerky flights from under a stone or the shelving bank, uttering his short

THE COMMON SANDPIPER.

note, *Wheet-wheet-wheet*, as he skims along. This is the COMMON SANDPIPER (*Totanus hypoleucos*)—family, *Scolopacidæ*—more generally known in the south of England as the Summer Snipe (but it must not be confounded with the snipe—*Scolopax gallinago*). It has many other provincial names, as Weet-weet, from its call-note, or Willywicket, Sand-Lark, Sand-Snipe, from its liking to sandy shores, and Fiddler in some parts of Scotland. About Shrewsbury it is called the Shad-Bird, as it came with the shads, which formerly, before the obstruction of the weirs on the Severn, came up as far as this city. Dresser ("Birds of Europe") says it is a wary bird. We have not found it so, oftentimes getting very close to it before taking wing. The white breast makes it very distinguishable as it runs from stone to stone. Look through your glasses and you will find he has a light brown back and head, with a white streak over the eye; legs dark-blue slaty colour; breast and under-parts white; hence the specific name, *Hypoleucos*.

The bird is very active and lively, and when on the ground is in constant motion. When on the wing the flight is peculiar—very rapid, with quick, jerky, but regular beats of the wings, at the same time uttering its call-note, *Weet-weet*. It breeds chiefly in the north of England and in Scotland, occasionally in Sussex and Devon, but the birds we see in the south are those on their migration. Yarrell and others state that this bird can swim and dive well, although its feet are not adapted to that purpose.

THE WATER-WAGTAIL.

One of our most pleasant and cheery companions of the water-side, as he passes with his drooping flight and sharply-uttered call-note, *Chiz-zit, chiz-zit*, is the WATER-WAGTAIL, or PIED WAGTAIL (*Motacilla Yarelli*); family, *Motacillidæ*. It is known amongst the country people as the Dishwasher. This name, Johns ("British Birds and their Haunts") says, "comes from the fanciful similarity between beating the water with its tail as it runs over the water-lilies and the beating of the water by washerwomen,"—a very far-fetched similarity indeed. It is, however, in some parts of France called *Lavandière*.

There are four British species of wagtails to be found

in our river-side rambles—the pied, the white, the yellow, and the grey; but the pied or black-and-white species is the most familiar, as he runs skipping about, now along the sides of the stream, then on the mass of weeds, or running on the tops of the hatches—always moving, always rest-

THE WATER-WAGTAIL.

less, but always with an eye on your movements; bold, but wary. All his world is for him:—

> "I'm a water-wagtail bold;
> All around, and all you see,
> All the world was made for me."

Watch it searching for its food, peering under every stone or leaf; its quivering tail never quiet; pert and happy all the day long; now up in the air after one of the *Ephe-*

meridæ, now chasing an insect along the bank. What activity! what airy lightness!

You will need your glass to distinguish him from his cousin, the WHITE WAGTAIL (*Motacilla alba*), which, although a somewhat rarer species, is no doubt often overlooked or confounded with the pied, the difference being in the colour of the back—the pied being quite black, the white more of a slaty colour, head and neck only being black. Their habits and localities are the same.

The COMMON YELLOW WAGTAIL (*Motacilla Raii*), of the same order and family as the last, is a regular summer

FIG. 1.—HEAD OF PIED WAGTAIL.

FIG. 2.—HEAD OF WHITE WAGTAIL.

FIG. 3.—HEAD OF YELLOW WAGTAIL.

FIG. 4.—HEAD OF GREY WAGTAIL.

visitant, and although often seen close to our rivers and streams, is not such a close frequenter of the water as the Dishwasher, but is partial to corn-fields and downs. It has the provincial name of the Barley-Bird, from arriving during the time of the Lent corn-sowing. From its delicate, fairy form and graceful movements, it is called in Spain *Pepita*, and on the Riviera *Ballarina*, or the Dancing-Girl. Its brilliant yellowy plumage attracts immediate attention; but if required to be seen well the binocular must be used, as the bird is not so familiar as his cousin above. Look at his beautiful yellow-green back and head, with the yellow

THE WATER-WAGTAIL.

streak over the eye; his bright yellow breast and underparts, with black bill and legs; and notice his long hindclaw, which in the water-wagtail is much shorter. The yellow wagtail is particularly fond of frequenting the meadows adjacent to water where cattle are grazing, knowing well that there flies do congregate; but he is also very partial to the larvæ of water-insects, and constantly alights on the *débris* of the reeds and rushes and water-plants which accumulate at mill-heads and hatchways. Its note is like the words "chit up"—shrill and sharp.

The GREY WAGTAIL (*Motacilla sulphurea*), which, according to Seebohm, "confines itself entirely to the water-side, may be distinguished from all other British wagtails by its uniting the characters of a grey back with a green rump and upper tail-coverts," and also by its black throat and chin, bound by a line of white. This bird has also a long hind-claw.

NEST OF THE REED WARBLER.

CHAPTER IV.

THE SWIFT.

WHEN Yarrell wrote his "History of British Birds," some forty-seven years ago, both swifts and swallows were placed in the natural order *Insessores Fissirostres*, and in the family *Hirundinides*. Since then ornithologists have placed the swallow, martin, and sand-martin in the order *Passeres;* family, *Hirundinidæ;* and the swifts are removed to the order *Picaridæ*—*i.e.*, with the nightjars and cuckoos; and with a family name to themselves—*Cypselidæ*. It is not necessary here to enter into a lengthened detail for this separation; but the editor of the fourth edition of Yarrell says:— "The characters which distinguish the swifts from the swallows are, even on a slight examination, so well marked and decisive that it is curious their important bearing on classification was not sooner recognised. Though so like swallows in much of their external appearance and in many of their habits, swifts have scarcely any part of their structure which is not formed on a different plan." He goes on to say that, "except a somewhat remote connection with the *Caprimulgidæ* (nightjars), the only true allies of the *Cypselidæ* are the humming-birds (*Trochilidæ*)."

These birds of the air appear never to rest, at one moment winging their rapid flight close to one, at another far, far away; now rising high up in the air above, now skimming the surface of the water searching for their prey, and snatching the duns and other *Ephemeridæ* almost out of the jaws of the lusty trout or lazy chub—always on the wing, from "earliest dawn till latest eve."

THE SWIFT.

THE SWIFT.

The Swift (*Cypselus Apus*) is easily distinguishable from its companions, the swallows and martins, by its greater size, its dark colour, and its long and powerful wings. See how these birds gyrate through the air—at one time so far above us as to be scarcely visible. See them hunting for flies in mid-air, at the same time having fine fun amongst themselves, chasing each other and shrieking. The rapidity of their flight is amazing. You may have been watching them one moment, and in the next there is not one to be seen—all have disappeared; and then, before you have done wondering where they could have gone to, they are again all around, flying here, there, and everywhere.

The swift is one of the latest birds to go to roost—long after the gloaming is passed his loud, shrieking whistle or scream is constantly heard whilst he is chasing his mate high up in the air; and then quite suddenly all is still.

The swift does not make a long stay with us: arriving about the end of April, it seeks more sunny climes than ours early in August. Stray ones are occasionally seen as late as October.

Gilbert White asks the question, "Where do swifts go when they leave us?"

> "Amusive birds! say where your hid retreat
> When frost rages and the tempests beat;
> Whence your return, by such nice instinct led,
> When spring's soft season lifts her blooming head?"

Most of our swifts come from and return to Africa, as is now well ascertained; but White, in his day, and others as well, believed that many swifts, swallows, and martins hybernated in holes in the banks of rivers, under the water, or elsewhere.

Gilbert White was the first naturalist to draw attention to the peculiarity of the foot of the swift, carrying all toes forwards. The least toe, which should be the back toe, consists of one bone only; the other three of two each.

The so-called edible swallows' nests, so much sought after by the Chinese to make into soup, are the nests of one of the swift family, and not of the swallow.

The swift builds under the eaves of houses, in church towers, and old barns; lays from two to four eggs of a dull white. They are very fond of the old thatched houses in country villages, and a most engaging sight is to watch their movements aloft; but it is no less interesting to behold some half-dozen birds racing, as they do within a few feet of the ground, through the narrow lanes of a small country village, uttering their squeaky note, *Smee-ee*, *smee-ee*:—

> "With sooty wing
> The shrill swift down the street before him swept."

THE SWALLOW.

In some countries it goes by the name "Screech" and "Deviling." There is an old notion that the swift cannot rise from the ground:—

> "This is my charter for the boundless skies,
> Stoop not to earth on pain no more to rise."

However, he is able to do this only with some difficulty on account of his long wings.

Swifts live entirely on insects of every kind. When the various hatches of the *Ephemeridæ* take place these birds frequent the river-side and devour vast quantities of May-flies, both of those dancing in the air or rising from the water.

THE SWALLOW.

The SWALLOW (*Hirundo rustica*, family, *Hirundinidæ*); CHIMNEY SWALLOW.

Sir Humphry Davy says in his "Salmonia:"—"I delight in this living landscape. The swallow is one of my favourite birds, and a rival of the nightingale, for he cheers my sense of hearing; he is the glad prophet of the year, the harbinger of the best season; he lives the life of enjoyment amongst the loveliest forms of nature; winter is unknown to him, and he leaves the green meadows of England in autumn for the myrtle and orange-groves of Italy, and for the palms of Africa. He has always objects of pursuit, and his success is sure; even the beings selected for his prey are poetical, beautiful, and transient. The friend of man, and, with the stork and ibis, may be regarded as a sacred bird."

There is an old Roman legend that swallows are the embodied spirits of dead children revisiting their homes. What a pleasant idea, and how it inclines us to love the bird still more! Ovid, in his "Metamorphoses," tells a terrible story of the sorrows of Procne, and of her being transformed into a swallow. The swallow was supposed to use the wild celandine plant to give sight to their young; hence the name of the bird in Greek is χελιδων, and the common name, swallow-herb for the celandine.

Willoughby, in his "Ornithology," gives a most wonderful account of the use of this bird for medicinal purposes, so that we may look upon it as a kind of "heal all." He quotes from Schrœder:—

1. "Swallows entire are a specific remedy for the falling sickness, dimness of sight, bleared eyes; for this their ashes are to be mixed with honey and applied as an ointment. They cure also the *squinancy* and inflammation of the uvula (pin of the mouth), being eaten, and their ashes taken inwardly.

2. "A swallow's head is good for the falling sickness, and to strengthen the memory. Some eat it against the quartan ague.

3. "Some will have the blood to be specific for the eyes, and they prefer that which is drawn from under the left wing.

4. "There is a stone found sometimes, though seldom, in the stomach of some of the young swallows called *chelidonius*, of the bigness of a lentil or pea. This, bound to the

arm or hung round the neck, helps the falling sickness in children.

5. "The nest applied gives relief to the squinancy (quinsey), heals the redness of the eyes, and is good for the biting of a viper or adder.

6. "The dung heats very much; the chief use is against the bitings of a mad dog, taken outwardly and inwardly, in colic, and in nephritic pains."

Then is given "an approved medicine" for the falling sickness (epilepsy). Take one hundred swallows ("I suppose," he writes, "there is some mistake, and that one quarter of this number will suffice"), one ounce of castoreum, one ounce of perony root, so much white wine as will suffice to distil altogether, and give the patient to drink three drachms every morning fasting.

Willoughby also says: "This bird is the spring's herald, being not seen throughout all Europe in wintertime, whence the proverb common in almost all languages, 'One swallow makes not a spring.'"

The earlier naturalists believed in the hybernation of swallows. Willoughby says it is more probable they go into Egypt and other hot countries; yet, later on, Gilbert White, as stated above, had a firm belief in their hybernation.

Moses Browne, in his angler's song:—

"Say canst thou tell where eels in winter hide?
Or where the swallows, vagrant race, reside?"

Wotton writes:—

"The swallow,
Lyre-like, attunes the sultry summer hours;
When chilling winter comes she torpid feels,
And fabricates her house amidst a tree."

Thomson had the same belief:—

"Clinging in clusters
Beneath the mouldering bark, or where,
Unpierced by frost, the cavern sweats."

We know better nowadays. The swallow comes to us

early in April, leaving us again in September or October (some late broods even remain till November) for the sunny climes of Africa, their migration having been successfully traced by way of the Mediterranean to that continent. What is more pleasant—

"Watching the swallow o'er the daisies flit,"

to contemplate on the wonderful power of wing which enables this bird for hour after hour, high up in the blue ether, or skimming the surface of river or meadow, to search for its daily food?

If swallows fly low, country people say rain is coming. A fall of temperature has driven the insects from the upper air downwards; the birds naturally follow their food.

The following curious account of a surgical operation performed by a parent swallow on one of its young is almost too marvellous. It is taken from the *Naturalist's World* for June 1886:—

"THE SWALLOW AS A SURGEON.—Dr. Walter F. Morgan, of Leavenworth, Kan., sends to the *Medical Record* this curious account of what may be called aviarian surgery, related to him in 1876 by the late Joseph O'Brien, Esq., of Cleveland, O.:—'On going into his barn Mr. O'Brien discovered a swallow's nest; and being a natural observer and lover of animals, he climbed to the nest, and found in it two young swallows, one being smaller and less vigorous than the other, and having a slighter covering of feathers. Upon taking the young bird in his hand, he was astonished to find one of its legs very thoroughly bandaged with horsehairs. Having carefully removed the hairs one by one, he was still more astonished to find that the nestling's leg was broken. Mr. O'Brien carefully replaced the bird in its nest, and resolved to await further developments. Upon visiting the patient the next day, the leg was again found bandaged as before. The bird-surgeon was not again interfered with, and the case being kept under observation, in about two weeks it was found that the hairs were being cautiously removed, only a few each day; and finally, when all were taken off, the callus was distinctly felt, and the

THE MARTIN.

union of bone evidently perfect, as the bird was able to fly off with its mates. Such instances may seem incredible to those not yet prepared to fully accept the axiom of the scientists, viz., that the intelligence of animals differs from that of man only in degree, and not in kind.'"

THE MARTIN.

The MARTIN (*Hirundo rustica*), resembling somewhat in its flight and action the chimney swallow. The two birds

THE MARTIN.

can be easily distinguished — the swallow by its long forked tail, red throat, and larger size; the martin by its shorter tail, white rump, white under-parts, and white-feathered legs and feet. The martin is perhaps a greater favourite, and is more familiar from its building its nest under our eaves or in the corners of our windows.

The martin delights to return every year to its old haunts, and either builds or repairs its old nest in the same corner year after year. Fearless of man, from their complete immunity from persecution, they mind not where they build, and a story is told by Mr. Benson (see Dresser's "Birds of Europe") of a pair of martins building their nest close to the wheel-house of the steamer *Orn*, which plied regularly on the river between Carlstad and Lychan in Sweden, the nest being only a foot or two above the water, and there they hatched their young. The birds when incubating travelled with the steamer backwards and forwards; but when the young were hatched the parent birds took up their quarters at Carlstad, and accompanied the steamer half-way on her trip, meeting her again on the return journey at the same place. All of us have seen how the sparrow will at times take the nest of the martin for his abiding-place, and what fights occur to regain possession; and there is a story told that, indignant at the outrage a pair of martins had received from this intruder, they deliberately clayed him up.

The martin from all time has been considered a sacred bird. Drayton goes so far as to say that primitive man made their huts of clay "learn'd from the martin;" and Shakespeare, in "Macbeth:"—

"This guest of summer,
The temple-haunting martlet."

Dryden calls it—

"A church-begot and church-believing bird."

When one sees these birds in all their happiness of love and nest-building, one recalls the lines in Wotton's beautiful sonnet, "On a Spring Day:"—

"Already were the eaves possest
With the swift pilgrim's daubed nest."

The velocity with which these birds fly was exemplified by an experiment recently made in Ireland with a view to ascertain this. On July 12 a house-martin (*Chelidon urbica*) was taken from a nest which contained young, under the eaves of Lowry's Hotel at Tubercurry, County Sligo, and after being placed in a cage, was conveyed ten miles away to Ballymote, where at 10.30 A.M. it was liberated. The nest was watched, and at 10.43 A.M. the bird returned, having accomplished the ten miles in twelve minutes, a rate of speed equal to fifty miles an hour.

An idea prevails amongst clean and tidy housewives that the martin brings lice to the houses, and therefore they often destroy the nest on that account. It is true that a kind of louse is found in the nests of the swallow and martin, but it is quite another species to that which attacks the human body, and is perfectly harmless. They swarm in the nests and on the birds, but can live nowhere else.

Another of these birds of the air, the SAND-MARTIN (*Hirundo riparia*), arrives in this country earlier than the two preceding, but is the smallest of the *Hirundinidæ*, and in early spring may be seen in considerable numbers hunting for its food up and down the river-side and over the surface of the water. It is easily distinguished by the mouse-coloured plumage of its upper-part. It breeds in holes in high sand-banks, in railway cuttings when through sand. In other ways its habits are much the same as the other members of the family. The sand-martin usually leaves us early—about the beginning or middle of September; but should the second broods be late, or any great change in the temperature take place, they are very likely to perish. We have been told, upon excellent authority, that in some seasons, when a severe frost has occurred early in September, great quantities of the young birds have been found dead floating on the surface of the Thames.

CHAPTER V.

THE LARK.

No bird is better known or more appreciated than the SKYLARK (*Alauda arvensis*; family, *Alaudidæ*). Its presence, either when standing on the top of a molehill among the flowers, raising its crest and scrutinising our movements, or when carolling in the air above us, imparts a sense of exhilaration not produced by the voice of any other of our songsters. Its glossy, light-brown back and speckled breast, with his long hind-claw—as Drayton says, "The lark with his long toe"—and full eye, is readily distinguishable when pert and erect on a clod; but when alarmed it can crouch and hide itself very effectually, and will not take wing till almost trodden on, particularly in the breeding season.

One of the earliest birds. The skylark often commences its song at break of day; hence the old saying, "Up with the lark." Chaucer calls it—

"The merrye larke, the messenger of day."

And we all know Shakespeare's lines from "Cymbeline:"—

"Hark, hark! the lark at heaven's gate sings,
And Phœbus 'gins to rise."

Elliott calls it the "Bird of the Sun:"—

"The cloud of the rain is beneath thee, thou singest
Palaced in glory; but morn hath begun
A dark day for man, while the sunbeam thou wingest,
Bird of the Sun, Bird of the Sun."

As regards the skylark's song, careful observers have

noticed that in the beginning of the season it seldom lasts more than two minutes, and in the full tide of spring, when

THE SKYLARK.

soaring up out of sight, never longer than a quarter of an hour. Swainson (" Provincial Names of British Birds ") gives the following lines to illustrate successfully the skylark's note—

> " La gentille alouette avec son tirelire,
> Tirelire, relire et tirelirant, tire
> Vers la voûte du ciel, puis, son vol en ce lieu
> Vire, et semble nous dire. Adieu, adieu, adieu ! "

and in noticing the difference in the song of this bird whilst on the ascent and descent, says : "A close observer of its habits has said, with reference to this, that ' the notes in the former case are of a gushing impatience, hurried out, as it were, from an excessive overflow of melody, which becomes gradually modulated, until the bird attains an elevation, when, as if satisfied with its efforts, it sinks gradually towards the earth with a sadder and more subdued strain.'" Dante has noticed this variation in his *Divina Commedia* :—

> " Like the lark,
> That, warbling in the air, expatiates long,
> Then, thrilling out her last sweet melody,
> Drops, satiate with the sweetness."

There is scarcely a British poet, from the earliest to the latest times, who has not sung the praises of the skylark ; and, as Professor Newton truly says, a volume might be filled with extracts describing or alluding to its marvellous power of song. All lovers of nature should remember Bloomfield's beautiful lines in his "Farmer's Boy." But a still more beautiful allusion to this bird, not so well known, was written by the Rev. Hart Milman on an incident he saw whilst reading the burial-service over Mrs. Lockhart, the daughter of Sir Walter Scott :—

> "Over that solemn pageant mute and dark,
> Where in the grave we laid to rest
> Heaven's latest, not least welcome guest,
> What didst thou on the wing, thou jocund lark,
> Hovering in unrebuked glee
> And carolling above that mournful company?

O thou light-loving and melodious bird,
  At every sad and solemn fall
  Of mine own voice, each interval
In the soul-elevating prayer, I heard
  Thy quivering descant full and clear,
  Discord not inharmonious to the ear!

We laid her there, the minstrel's darling child;
  Seem'd it then meet that, borne away
  From the close city's dubious day,
Her dirge should be thy native wood-note wild;
  Nursed upon Nature's lap, her sleep
  Should be where birds may sing and dewy flowerets weep."

From its blithe, joyous song, the skylark is often kept in captivity. Gay, in Epistle IV., makes the bird relate how the advantages of his song doom him to captivity and misery:—

"For what advantage are these gifts to me?
  My song confines me to the wiry cage;
  My flight provokes the falcon's fatal rage!"

There is one consolation to the poor caged prisoner. Could he but know it, he gives solace and joy to many a poor bedridden sufferer, and to others in our crowded cities, in hours of toil and pain; but when we see it trampling on its bit of turf in its bow-windowed cage, with head upraised and fluttering wings, we realise that it has indeed lost its liberty, and no longer

"Singing still dost soar, and soaring ever singest."

It is remarkable how the skylark easily distinguishes its enemies amongst the *Falconidæ*. Professor Newton says: "The appearance of a merlin will cause the sudden cessation of the song—at whatever height the performer may be, his wings are closed, and he drops to the earth like a falling stone. The kestrel, however, is treated with indifference, and in the presence of a sparrow-hawk the skylark knows that safety is to be sought aloft."

The skylark's nest is always placed on the ground, and generally contains three to five eggs of a French white colour, freckled with brownish blotches. Grahame, in his

"Birds of Scotland," contrasts the low situation of the nest with the bird's aerial flight:—

> "Thou simple bird,
> Of all the vocal choir, dwellest in a home
> The humblest; yet thy morning song ascends
> Nearest to heaven."

Two other larks, but of a different family, *Motacillidæ*, are often seen and heard not far from the river-side.

The TREE PIPIT or PIPIT LARK (*Anthus Arboreus*) is common enough, and generally visits the enclosed districts.

THE TREE PIPIT.

It arrives in this country in April, and departs again in September. The male is generally to be seen sitting on some outer branch of a neighbouring tree or bush, and constantly rises up in the air, singing as he goes, and then returns with quivering wing and outstretched tail to the place he started from, or at times down to the ground, always ending his short song with a long-drawn *Tsee-a, tsce-a*, oftentimes repeated. It is much the colour of a lark, only darker.

The MEADOW PIPIT or TITLARK (*Anthus Pratensis*) is a resident all the year, but not so common among the southern meadows; but farther north, amongst the moors and open districts, along the rocky sides of mountain streams, it may be constantly seen sitting on heather-stems, bushes, and walls, singing not so loud as the tree pipit, but with the same motions.

THE MEADOW PIPIT.

It is similar in appearance to the tree pipit, but somewhat larger, has a longer hind-claw, and is more speckled on the breast.

THE STARLING.

The STARLING (*Sternus vulgaris*)—family, *Sturnidæ*—is another bird constantly found in the water-meadows and by the river-side, and a very pert and pleasant bird it is, restless and busy from morning till night, seeking for worms or other articles on its dietetic list. These birds, in search of their food, appear to force their beaks into the moist earth and dig out beetles and worms without number.

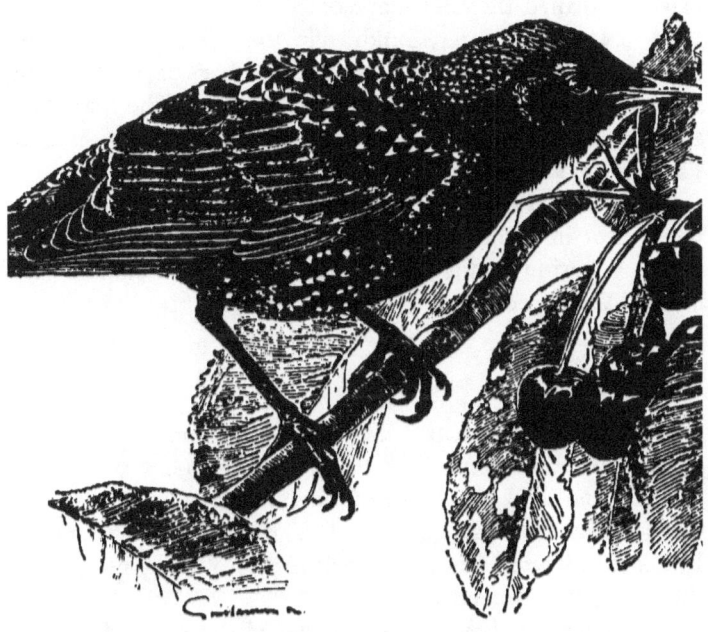

THE STARLING.

Watch them among a flock of sheep and cows, perched on the wool or backs of the cattle, searching for the ticks and bots, fearless and confiding. The angler constantly disturbs starlings from the high flags close to the river. We suspect they come after the caterpillars and moths hiding amongst the reeds. The bird is also particularly partial to cherries. The starling, both male and female, has a beautiful plumage. At a short distance they look black, without any other colour, but if you put your glass upon them when near, you will find they shine with splendid iridescent colours of purple and green. The young bird is a dark greyish-brown, and in earlier times was supposed to be a different species, and described as the "solitary thrush." The wing-feathers of the young starling are much used for making certain artificial flies.

The bird is often called the Stare, which is supposed to be derived from the Anglo-Saxon word *staer*. It is also in some counties, from its habit of perching on the backs of sheep, called the Shepstare or Shepster.

It is an early nester, building in the holes of trees, often taking the hole made by the woodpecker, or ousting the sparrow from under the eaves of buildings. The male bird assiduously attends its mate when she is sitting, and, as Fawkes says, "whistles curious notes of love."

It often imitates the voice of other birds, particularly when kept in confinement. When disturbed or angry, it makes a curious croaking noise, somewhat similar to the note of the corncrake.

THE CUCKOO.

In some parts of England, the CUCKOO (*Cuculus canorus*), of the family *Cuculidæ*, has the provincial name of Gawky. In Scotland and the north of England it is known as the Gowk. In the last edition of Yarrel's "British Birds," Professor Newton has gone back to the old way of spelling the name, viz., Cuckow.

The cuckoo's first notes are hailed by all, and especially by springtide fishermen, as a certain sign that dreary winter is over, to be replaced by balmy breezes, flowery meads, and songs of birds, making "all nature gay."

We should sadly miss this bird if ever he became a stranger among us; when we hear him in the distant wood, or see him as he flies across the river close by us, we know our time is come again:—

> "The merry cuckoo, messenger of spring,
> His trumpet shrill has thrice already sounded."

THE CUCKOO.

Cuckoo folk-lore abounds in fabulous stories: how it changes, according to Aristotle and Pliny, into a bird of prey—a belief still extant; some of the old gamekeepers in the north of England affirm that the bird always changes into a hawk in the winter. In Cambridgeshire the bird is a cuckoo for three months and a hawk for nine. In Bohemia the same belief exists. In Switzerland they go so far as to say that the cuckoo of one year will be a young eagle the next. In North Germany it becomes a

THE CUCKOO.

sparrow-hawk after St. John's Day, and the same legend is to be found in many parts of France. In Normandy—

> "Entre Juin et Juillet,
> Le coucou devient emouchet."

Pliny says: "If the cuckoo is wrapped in a Hare's skin and applied to a patient, it will produce sleep." Rondoletius tells us that its ashes are good for disorders of the stomach.

The cuckoo was supposed to be an infallible remedy against fleas. If, when first heard, the hearer circumscribed his right foot and dug up the earth around, not a flea would be found wherever that earth was scattered.

Its time of appearance has also produced innumerable rhymes. One of the oldest ballads in the English language is about the coming of the cuckoo:—

> "Sumer is icumen in,
>   Shude sing cucu.
> Groweth sed and bloweth med,
> And springeth the wode nu.
>   Sing cucu.
> Awe (ewe) bleteth after lamb,
>   Showth (loweth) after calve, Cu,
> Butter sterteth,
> Bucke verteth,
>   Music sing cucu.
>   Cucu-cucu."

Another says:—

> "In March he leaves his perch,
> In April come he will,
> In May he sings all day,
> In June he changes his tune,
> In July he is ready to fly,
> Come August go he must,
> In September you'll him remember,
> But October he'll never get over."

In Derbyshire and in the north of England they have the following:—

> "In April cuckoo sings his lay,
> In May he sings both night and day,
> In June she loses her sweet strain,
> In July she is off again."

About the middle of June the cuckoo's note is much changed; instead of the loud, plain *Cuckoo-cuckoo*, we hear a double first note, *Cu-cuck-oo*. This was noticed by John Heywood, who wrote in the latter part of the sixteenth century or the beginning of the seventeenth:—

> "In April, the koocoo can sing her song by rote,
> In June of tune she cannot sing a note;
> At first koo-coo—koo-coo, sing still she can do,
> At last kooke-kooke-kooke, six kookes to one koo."

Among other bad habits, she is accused of sucking eggs. Francis Quarles says:—

> "The idle cuckoo, having made a feast
> Of sparrows' eggs, layes down her own i' the nest."

Rennie says that the cuckoo, by placing her eggs in the other birds' nests, annually destroys 3,500,000 of their eggs. The old nursery song says the male bird does so for a special purpose:—

> "He sucks little birds' eggs
> To make his voice clear."

In Denmark the reason why this bird makes no nest is thus accounted for:—"When, in early spring, according to Mr. Horace Marryat ('Jutland and the Danish Isles'), the voice of the cuckoo is first heard in the woods every village girl kisses her hand and asks the question, 'Cuckoo, cuckoo, when shall I be married?' And the old folks, borne down with age and rheumatism, inquire, 'Cuckoo, when shall I be released from this world's care?' The bird, in answer, continues singing *Cuckoo* as many times as years will elapse before the object of their desires will come to pass. But as some old people live to an advanced age, and many girls die old maids, the poor bird has so much to do in answering the questions put to her, that the building season goes by and she has no time to make her nest, but lays her eggs in that of the hedge-sparrow" (Swainson, "Provincial Names of British Birds").

The cuckoo is also invested with all kinds of power for good or evil, luck or ill-luck. Macgillivray says: "In the Hebrides, and other parts of the west of Scotland, if the cuckoo is first heard by one who has not broken his fast, some misfortune is sure to occur. The same sort of thing is believed also in France; and in Germany if a person hears a cuckoo before his first meal, it entails hunger for a year; and to hear the cuckoo's first note when in bed is woe indeed."

Where the bird went to was a mystery in olden times, and the old story of hybernation during the winter is told by Aristotle and later writers; how, in burning an old hollow log in winter, cuckoos would suddenly appear. Thomas Carew writes :—

> "But the warm sun thaws the benumbed earth,
> And makes it tender, gives a sacred birth
> To the dead swallow, wakes in hollow tree
> The drowsy cuckoo and the humble bee."

The usual time for the arrival of the cuckoo is about the middle of April. The male birds arrive some little time before the females, and commence at once their well-known call. They are more numerous than the females. These have an entirely different note, more like that of the dabchick, a kind of running whistle often repeated.

Dresser says : "The note of the male is the well-known call which is generally heard, and consists of two syllables, *Uh-uh* rather than *Ku-ku*, which when the bird is greatly excited is rendered *Ku-ku-ku;* and besides this, it utters a peculiar harsh note which somewhat resembles the syllables *Quawawa* or *Haghaghaghag*. The female, on the contrary, has quite a different call, a sort of laughing note uttered very quickly, like the syllables *Jekikickick* or *Quickwick-wick*, which it preludes with a low harsh sound."

From a number of careful observations which have been made in this country and abroad, it is now ascertained that the cuckoo usually lays her egg on the ground, and then with her beak deposits it in the nest she has chosen for that purpose. The egg has so often been found in nests placed in such a position that it would be impossible for the bird to have laid it there, and where she could only have put it with her beak. Macgillivray recorded in 1838 an instance of the bird so placing her egg in a titlark's nest; and Herr Adolf Müller (see Yarrell's fourth edition), a forester in Darmstadt, notifies that he distinctly saw through a telescope a cuckoo lay her egg on the bank, and then take it in her bill and deposit it in a wagtail's nest close at hand, where he immediately afterwards found it; and many other instances have been recorded. The egg of the cuckoo is often extremely similar to the egg of the bird whose nest she has chosen. Seebohm gives no less than fifteen examples of the cuckoo's egg—not one alike. At a meeting of the Royal Society in 1884 a German naturalist exhibited about thirty nests of different birds in which the cuckoo had laid or deposited its egg, and in many the cuckoo's egg was almost similar in colour to the egg of the bird which had built the nest. The cuckoo's egg has been recorded as found in the nests of no less than ninety-three species of birds (thirty-seven in this country); but the most usual nests are the hedge-sparrow's, reed-warbler's, pied wagtail's, and meadow pipit's.

We all know how the young cuckoo, by means of a depression in its back, is able to hoist up and pitch his foster brothers and sisters out of the nest, there to die of cold and starvation. It is indeed a strange provision of nature that, although the parents will feed and guard their big foster-child with the greatest solicitude, they will take no heed of their own callow young lying on the grass close by.

The adult bird, as we see him, is very handsome—the dark bluish-grey head, back, and throat, the barred black and white breast, under-parts, and tail, bright yellow eye, and yellow legs. The young bird, which we see in July and August, is quite different—the head and neck a brownish-grey, with a white spot at the back of the neck; back, wings, and tail a clove-brown, the feathers tipped with white; breast and under-parts a dirty white, with brown bars; eyes brown. There is no difference in the plumage of the sexes. The female is somewhat smaller than the male bird.

Cuckoos do not pair. The old birds generally migrate south in August; the young birds later in September. The various notices of the cuckoo being heard in this country in December or early in the year are simply caused by imitative boys; in all these accounts the bird is said to have been heard, but is never seen.

The greater number of our cuckoos migrate to Africa, and many residents in Algiers hear the familiar note about the time of the blossoming of the hawthorn. Mr. Blandford has heard the cuckoo amongst the Baluchistan Hills in Northern India in the months of February and March. We combine its advent with April showers and May flowers, and when listening to its well-known voice call to memory Logan's beautiful lines :—

> "Hail, beauteous stranger of the grove,
>   Thou messenger of spring!
> Now Heaven repairs thy rural seat,
>   And woods thy welcome sing ;
> What time the daisy seeks the green,
>   Thy certain voice we hear.
> Hast thou a star to guide thy path,
>   Or mark the rolling year?
> Delightful visitant! with thee
>   I hail the time of flowers,
> And hear the sound of music sweet
>   From birds among the bowers.
> The schoolboy wandering thro' the wood
>   To pull the primrose gay,
> Starts, the new voice of spring to hear,
>   And imitates thy lay.
>   .   .   .   .   .
> Sweet bird, thy bower is ever green,
>   Thy sky is ever clear ;
> Thou hast no sorrow in thy song,
>   No winter in thy year."
>   .   .   .   .   .

THE SONG-THRUSH.

In early spring, sometimes indeed in the winter months, if free from frost, high up among the branches of the elms and poplars, the SONG-THRUSH (*Turdus musicus*), of the family *Turdidæ*, pours forth its notes "from daybreak to the silence of the groves." His sombre plumage conceals him from view; but his voice (softer in February, but becoming much more powerful as spring sets in) rings through the welkin, and can be heard at a considerable distance :—

> "His note so clear, so high,
> He drowns each feather'd minstrel of the sky."

THE SONG-THRUSH.   93

The bird is not at all shy, and to watch it on a lawn after rain is very amusing; it runs up to a worm cast, eyes it with a peculiar look, begins pecking at it, and stamping with its legs. The poor worm, alarmed, pokes out his head to see what can be the matter. Alas, poor worm!—he is gobbled up in a moment.

It is very fond of snails, and carries them to a big

THE SONG-THRUSH.

stone, there to break the shell and consume the contents. The so-called thrush's dining-table may often be seen as we pass through a copse on our way to the river-side.

The nest is beautifully made, and lined with clay. The eggs are generally five, of a light blue speckled with black. The female sits very close, almost allowing you to touch her before leaving the nest. She watches you narrowly with her dark, lustrous eye. Andrew Marvel says :—

> "And through the hazels thick, espy
> The hatching thrussel's shining eye."

Burns also :—

> "An' she's twa glancing sparklin' een."

Often on returning late from the river our ears are greeted with the song of this bird singing the "drowsy day to rest."

The thrush sings almost all the year round, from the earliest spring till "chill October."

The late Frank Buckland, in *Land and Water*, gives the following very characteristic wording of the thrush's song :—

> "Knee-deep, knee-deep, knee-deep ;
> Cherry-du, cherry-du, cherry-du, cherry ;
> White hat, white hat ;
> Pretty joey, pretty joey, pretty joey."

Macgillivray gives another version :—

> "Qui qui qui, kween, quip,
> Tiurru, tiurru, chipiwi ;
> Too-tee too-tee, chiu-choo,
> Chirri, chirri-chooee,
> Quiu-qui, qui."

Harting ("Ornithology of Shakespeare") says : "It must be admitted by all who have paid particular attention to the song of the thrush, that this is a wonderful imitation as far as words can express notes, and this is rendered more apparent if we endeavour to pronounce the words by whistling."

THE BLACKBIRD.

The BLACKBIRD (*Turdus merula*), of the same family as the thrush.

> "The ousel cock, so black of hue,
> With orange tawny bill."

THE BLACKBIRD.

The favourite of our woods and gardens (perhaps not of the gardener!), how black his plumage, how orange his beak, and how the copse rings with his alarm-note when a cat or a fox, an owl or a hawk, or any other enemy to his race, crosses his path or nears his nest; but how mellow and sweet is his song as he tunes his dulcet pipe, morning and evening, to his mate sitting on her well-concealed nest in holly or ivy! The nest differs from the thrush's in being lined with fine bents or soft grass. The eggs, four to five, are greenish-grey ground, mottled with reddish spots.

In the early Jacobite days the blackbird was a symbol

THE BLACKBIRD.

of the cause, in allusion to Charles the Second's swarthy complexion. One of the stanzas of an old Scotch song runs thus:—

> "Once in fair England my blackbird did flourish,
>   He was the chief blackbird that in it did spring;
> Prime ladies of honour his person did nourish,
>   Because he was the true son of a king."

The old name Merle, by which this bird was formerly known, is derived from the Latin *merula*.

Drayton calls it "the mirthful merle."

"The woosel and the throstle cock chief music of our May."

The woosel is evidently a corruption of ousel.

In some parts of Germany the backbird is kept in a cage and hung outside the house as a protection against lightning. Swainson relates a legend of St. Kevin, when that saint, praying in the Temple of the Rock at Glendalough, with outstretched hand, a blackbird laid its eggs in the palm, and the compassionate saint never removed his hand until the eggs were hatched!!

Montgomery well describes the blackbird's notes:—

> "Then, with simple swell and full,
> Breaking beautiful through all,
> Let thy pan-like pipe repeat
> Few notes, but sweet."

THE NIGHTINGALE.

From the middle of April till leafy June, in the thickets around and about our southern rivers, we may often hear, but seldom see, the so-called queen of song, the NIGHTINGALE (*Daulias luscinia* family *Sylviidæ*); but whenever heard, its song arrests the attention and fascinates the ear of all; no bird sings forth its notes with greater force or with more thrilling cadences. Izaak Walton says: "She breathes such sweet music out of her little intermittent throat, that it might make mankind to think that miracles have not ceased."

John Hunter found that the muscles of the larynx of the cock nightingale were stronger and more fully developed than any other bird of its size.

The male bird arrives some fourteen days before the female, and when undisturbed, sings almost as much in the daytime as at night; occasionally one may be observed sitting on a bare branch in the brake, sending forth a volume of rich notes. George Gascoigne endeavoured to give them in words:—

> "Yet never hearde I such another note—
> Terew, terew; and thus she gan to plaine
> Most piteously, which made my heart to grieve.
> Her second note was fy-fy, fy-fy-fy,

And that she did in pleasant use repeat
With sweet reports of heavenlie harmonie.
She showed great skil—for times of unisone,
Her jug-jug-jug in griefe had such a grace.
Then stinted she, as if her song was done,
And ere that past, not ful a furlonges space,
She 'gan again in melodie to melt ;
But one strange note I noted with the rest,
And that saith thus : Nemesis, Nemesis."

THE NIGHTINGALE.

Milton's beautiful lines in "Il Penseroso" are familiar to all :—

> " Sweet bird, that shunn'st the noise of folly,
> Most musical, most melancholy !
> Thee, chauntress, oft, the woods among,
> I woo, to hear thy even-song."

In Father Faber's beautiful poem, "The Cherwell Water-Lily," are the following touching lines on this bird:—

> "I heard the raptured nightingale
> Tell from yon elmy grove his tale
> Of jealousy and love—
> In thronging notes, that seemed to fall
> As faultless and as musical
> As angels' strains above;
> So sweet, they cast o'er all things round
> A spell of melody profound;
> They charmed the river in his flowing,
> They stayed the night-wind in his blowing,
> They lulled the lily to her rest
> Upon the Cherwell's heaving breast."

The legend of the nightingale always singing with its breast against a thorn arose from the idea that it nests and roosts in thorny places to avoid serpents, to which it was supposed to have a particular dread:—

> "Au printemps, doux et gracieux
> Le rossignol a pleine voix
> Donne louange au dieu des dieux,
> Tant qu'il faict retentir les boys.
> Peur du serpent il chante fort,
> Toute nuict et met sa poictrine
> Contre quelque poignante espine
> Qui le reveille quand il dort."

Shakespeare in "Lucrece" says: "The adder hisses where the sweet bird sings."

The nightingale has its mythological history, and a very sad one. Philomela, daughter of Pandion, despoiled by Tereus, fled to the woods, and there, transformed into a nightingale, with its breast pierced by a thorn, she pours forth expressions of her misery and woe to the silent listener of the night. Hence all the poets connect it with melancholy and grief, also place the bird in the feminine gender. The song is seldom or never heard after the young are hatched. It is hardly necessary to say that it is the male bird alone which sings.

When the young are hatched—early in June as a rule—the song ceases, and gives place to a croaking note of alarm, the male bird joining with its mate to procure food for its young. The song of the nightingale, depending thus on the hatching of the young, may in some places be heard later than in others. Shakespeare gives another reason for the song ceasing:—

> "As Philomel in summer's front doth sing,
> And stops her pipe in growth of riper days;
> Not that the summer is less pleasant now
> Than when her mournful hymns did hush the night,
> But that wild music burdens every bough,
> And sweets grown common lose their dear delight."

The nightingale is dark russet-brown from head to tail, but shading to chestnut-brown on the lower portions. The under-parts are buffish-white, shading to grey; the eye hazel; the legs brown. The nest is placed close to or on the ground, generally in a depression, loosely constructed of dried oak, hawthorn, and other leaves, lined with fibrous leaves. The eggs, four or five, of a deep olive-brown. Part of the day she remains near the nest, ensconced among the low brambles, &c. The nightingale has been but rarely heard farther north than the city of York, or west than the line of the river Exe, and it has never been heard in Ireland.

THE WILLOW-WARBLER.

There are two or three little birds which creep about the bushes close to the water-side seeking their insect food. One of these is the WILLOW-WARBLER or WILLOW-WREN (*Phylloscopus trochilus*, or the Leaf-Searching Wren); family, *Sylviidæ*. He is a very restless, lively little fellow, working his way amongst the plants and bushes and dwarf willows, now and again uttering a few low but rather sweet notes, not taking any notice of you if you do not appear to notice him; but stop for a second or two and fix your eye on him and he rapidly disappears. The white throat and under-parts make it easily seen, and the green olive plumage of the back and white stripe over the eye distinguish it from another somewhat similar in plumage, but with a less defined white mark over the eye, and of a slightly more sombre green plumage, the CHIFF-CHAFF (*Phylloscopus collybita*), whose cheerful monotonous song of *Chiff-chaff, chiff-chiff chat*, often repeated, may be heard on the

THE WILLOW-WARBLER.

tops of the highest trees when the hazels are covered with their catkins and the sallows with their down. It is, with the exception of the gold-crest, one of our smallest migrant warblers, as well as one of the earliest to arrive. The

THE WHITE-THROAT.

back is olive-green, darker than the willow-wren. It has a small stripe of light colour behind the eye. The under-parts are white, shaded with grey. The bird leaves us again in the middle of October. In France they call this bird *Compteur d'Argent*.

THE WHITE-THROAT.

The WHITE-THROAT (*Sylvia cinerea, Curruca cinerea*), or the bird which hatches the cuckoo's egg, is of a reddish-

THE WHITE-THROAT.

brown colour on the back, the under-parts pale brownish-white tinged with rose-colour, the chin and throat being white—hence the name. Its manners differ also from the willow-warbler, often letting you know of its presence by the curious alarm-note, *Chzh-chzh;* and you may see him on the topmost twig of a bramble or the upper part of the flowering stem of the meadow-sweet, with head-feathers erected into a crest, his throat full and quivering with excitement, whilst he pours out a rather sweet but short song. Bloomfield says:—

> "The sporting white-throat, on some twig's end borne,
> Pour'd hymns to freedom and the rising morn."

Seebohm says: "In the early summer he is so full of music that sometimes, as he flies from hedge to hedge, he will soar up into the air above his line of flight and pour out his song like a pipit or lark. The bird is a fruit as well as an insect eater. Very fond of daddy-longlegs, as well as of currants and raspberries."

It has many provincial names. In some parts of Scotland it goes by the name of Blethering Tam. In England it is the Nettle-Creeper, Haytit, Billy Whitethroat, Great Peggy, &c., &c. Swainson says that "the name of Singing Sky-Rocket has been applied to it from its habit of rising quickly from time to time straight up in the air, singing all the time." In France it is called *Babillarde*.

THE WREN.

The WREN (*Troglodytes parvulus*), the little bird that lives in a hole—family, *Troglodytes*—is one of the most familiar of our bird companions. It matters not where you may be, whether by the river-side or on the garden-seat, if you remain on one spot for a very short time a wren will be sure to come and have a look at you. Many and many a time, when we have been "far from the madding crowd," in most out-of-the-way places, sitting quietly down either to rest, to sketch, or to change our flies, our attention would be drawn to something moving among the bushes—a little brown object would appear and as quickly disappear, and then again out it would come with its tail cocked up, peering at you with a pert yet confiding look, and treating you with a burst of his thrilling, quick, and bustling song. One is never lonely when a wren is near.

In many parts of England to kill a wren is tantamount to bringing down on the family of the murderer, poverty and misfortune, if not something worse.

In Ireland, however, this poor little bird was formerly

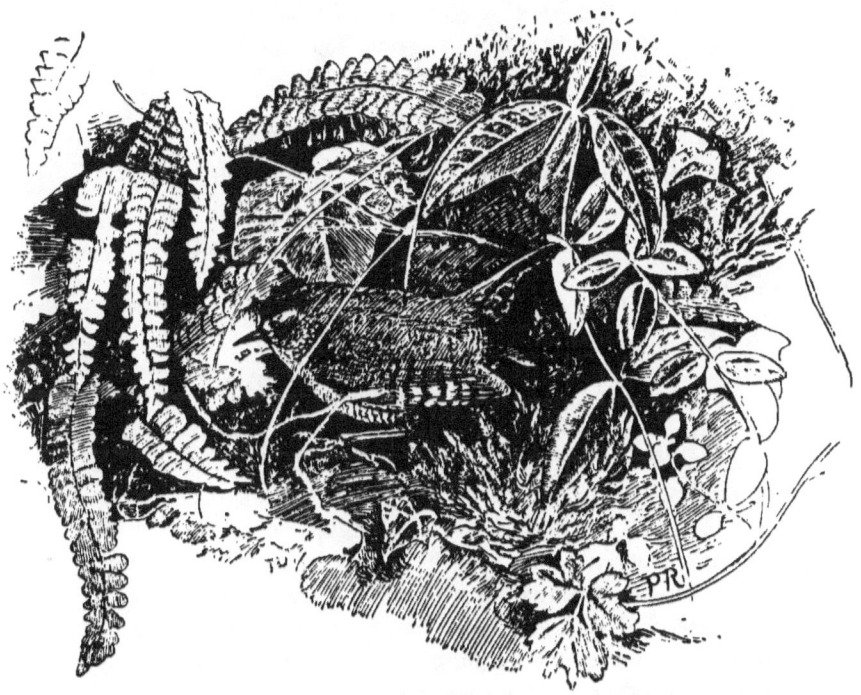

THE WREN.

subjected to severe persecution on Christmas Day, when the hunting of the wren took place—a barbarous custom, the origin for which is not well known. Mr. and Mrs. S. C. Hall, in their "Ireland, its Scenery and Character," say that when the Danes invaded Ireland the Irish were on the point of surprising their enemies during sleep, when a wren perched upon a drum awoke the sentinels just in time to save the army, whereupon the bird was denounced as a traitor, outlaw, &c., and condemned to be killed whenever met.

On St. Stephen's Day the hunting of the wren and then slinging the poor birds on a pole was a usual custom in the south of Ireland. The children with this pole went about soliciting alms and singing a song, part of which is here quoted from Mr. S. C. Hall's book :—

> "The wran, the wran, the king of all birds,
> St Stephen's Day was cot in the furze;
> Although he is little, his family's grate,
> Put your hand in your pocket and give us a trate.
> Sing holly, sing ivy, sing ivy, sing holly,
> A drop just to drink, it would drown melancholy."

The beautiful moss-domed nest of this little bird is often found in the roof of a summer-house, or in some ivy, or by the side of an old tree. Wordsworth says :—

> "Among the dwellings framed by birds
> In field or forest with nice care,
> Is none that with the little wren's
> In snugness may compare."

And Mr. J. Whitaker gives the following interesting account of these birds building their nests with the materials nearest at hand :—

"WRENS' NESTS.—Some time ago I described in the *Zoologist* a wren's nest which had been built in a straw-stack, and as the outside was entirely composed of straw, I remarked that these little birds assimilated their nests generally to the surrounding objects. Since then I have formed another opinion—namely, that they make their nests of the nearest available materials, which very often match the surroundings of the nest. I may mention the following nests which have come under my notice :—One in brown bracken, all outside of bracken; one in a wall over a bed of nettles, the outside being composed of pieces and leaves of nettles; one near a carpenter's shop, all outside of shavings; one in an arbour—here the nest was built in the side in some old heather, and was made of old grass which was used to stop up holes in the window of the arbour; the light-brown grass was very conspicuous against the dark heather. One in a beech-tree on the lawn was formed of new-mown grass from a heap below the branch on which it was placed."

THE REDBREAST.

The REDBREAST (*Erytheca rubicola*); family, *Sylviidæ*. This lively, familiar, but very pugnacious bird is generally distributed through the British Isles, and is a universal favourite with all classes. Its sprightly air, full, dark eye, and the great confidence it displays in its associations with man endear it to all, and the thought of killing a robin

THE REDBREAST.

in this country is considered about as sacrilegious as robbing a church. Not so, however, in other countries, especially in Italy, where you may see strings of robins and other small birds hung up for sale on market-days. The dish of *becca fici* served up at the *table d'hôte* is chiefly composed of robins. Browning says :—

> "A man may have an appetite enough
> For a whole dish of robins ready cooked."

With us the bird is so connected with our childhood and its stories (for who does not know "The Children in the Wood" or "Who Killed Cock Robin?") that to think of eating him would take away the appetite entirely.

Both male and female have the red breast, but the male plumage is generally richer and brighter. How this soft-billed bird lives through the hard winter months is a marvel, and this circumstance is probably the origin of the old couplet:—

> "The robin and the wren
> Are God A'mighty's cock and hen,"

as well as the supposed office of covering with leaves and moss the dead bodies of unburied mortals. Robert Herrick wrote an ode to robin redbreast, asking him to perform this rite upon him:—

> "Let thy last kindnesse be
> With leaves and mosse work for to cover me;
> And while the wood nymphs my cold corpse inter,
> Sing thou my dirge, sweet warbling chorister."

Shakespeare gives the bird the same kind office:—

> "The ruddock would
> With charitable bill (O bill, sore-shaming
> Those rich-left heirs, that let their fathers lie
> Without a monument!) bring thee all this;
> Yea, and furred moss besides, when flowers are none
> To winter-ground thy corse."—*Cymbeline*, Act iv. sc. ii.

The robin has a beautiful voice, and sings when all other birds are becoming silent. He loves to frequent the habitations of man, and we all know how fond he is of a church, often joining in sweet melody the hymns of praise poured forth by his human companions.

It is stated as a well-known fact that a robin frequently perched on one of the pinnacles of the organ in the Cathedral of Bristol and joined the music with its song.

Peter Pindar, in his ode to some robins in a country churchyard, calls them "wild tenants of the fane," and old Skelton, in his "Elegy on Philip-Sparrow," gives the robin

> "The requiem mass to sing,
> Softly warbelynge."

The bird has a number of familiar names, such as Bob Robin, Bobby, Robin Ruck, Ruddock, &c. In Sweden he is called *Tomi-Liden;* in Norway, *Pieter Ronsmad;* in Germany, *Thomas Gierdet.*

Wordsworth, in his sonnet, "The Redbreast and Butterfly," says:—

> "Art thou the Peter of Norway boors?
> Their Thomas in Finland,
> And Russia, far inland?
> The bird who, by some name or other,
> All men who know thee call their brother,
> The darling of children and men?"

There are many legends connected with this bird. In Scotland and in the north of England, as well as Germany, it is believed that if a robin is killed, one of the cows of the slayer will give bloody milk. M'Gregor ("Folk-Lore of West of Scotland") says:—"There is a popular saying that the robin has a drop of God's blood in its veins; therefore to kill or hurt it is a great sin."

> "No wanton boy disturbs its nest,
> Weasel nor wild cat will her young molest;
> All sacred deem the bird of ruddy breast."

In Bohemia the slayer of a robin will always suffer from shaking paralysis of the hands. In Brittany there is a legend that the robin owes its red breast from having in pity plucked a thorn from our Saviour's crown when hanging on the cross, and as a reward the bird is endowed with perpetual existence.

The robin is able to hold his own with other birds when, in winter, the scraps of the breakfast-table are thrown upon the lawn for their benefit. A friend writes:— "The first to come down is the robin, then the sparrows, then the chaffinches, then the thrushes, and then the blackbirds, and what they do is this. If only one sparrow comes, the robin flies at him and drives him off; if two sparrows come, he is still equal to the occasion; but if three come he is puzzled, and it is curious to see, what with his attention to the bread and to these interlopers, how perplexed he is; and if four come, he gives it up and flies away. If a chaffinch comes he seems to hesitate, but every feather becomes erect, and he makes himself as big as two. The chaffinch also does not like his appearance, and hesitates to approach the bread. Then if another chaffinch should come, the two set upon Master Robin and drive him off. But though I have often looked for it, I have never seen two birds of different species combine in the attack, though in the absence of the robin they have no apparent jealousy of each other, and all feed together, the robin being the only one amongst them which seems to take a line of his own."

The red breast appertains to both sexes. Their alarm-note is a sharp *Tick-tick*, but when the young are hatched and the nest is approached, this is changed to a short, wailing, plaintive pipe. The young, till the first moult, are speckled.

Rogers wrote the following epitaph on a robin redbreast:—

"Tread lightly here; for here 'tis said,
  When piping winds are hush'd around,
  A small note wakes from underground,
Where now his tiny bones are laid.
No more in lone and leafless groves,
  With ruffled wing and faded breast,
His friendless, homeless spirit roves;
  Gone to the world where birds are blest!
Where never cat glides o'er the green,
Or schoolboy's giant form is seen;
But love and joy, and smiling spring,
Inspire their little souls to sing."

THE BLACK-CAP.

Where the stream flows by gardens or shrubberies we often hear the song, and still more often the warning note, of the BLACK-CAP (*Sylvia atricapilla*), which amongst the *Sylviidæ* claims the second place as a songster (the nightingale being the first). If you should hear its alarm-note, *Sharr-sharr* (a sign he has seen you), it is worth while to rest a few minutes and listen for the song. The bird, hidden amongst the low bushes, will gradually make its way up to some rather bare spray, and there, if not disturbed, you will see it throw up its head, elevate its crest, enlarge the throat, and beginning with a few soft notes, gradually increase them till they become loud, joyous, and

THE BLACK-CAP.

prolonged, and then suddenly cease; but the tone through the whole is very sweet. Gilbert White, whose observations on birds are always of great interest, and very truthful, says: "The black-cap has, in common, a full, sweet, deep, loud, and wild pipe; yet that strain is of short continuance, and his motions are desultory; but when that bird sits calmly and engages in song in earnest, he pours forth very sweet and inward melody, and expresses great variety of soft and gentle modulations, superior perhaps to those of any of our warblers, the nightingale alone excepted." And, like the nightingale, the black-cap ceases to sing as soon as the young are hatched. But you have moved, and the bird has disappeared, and instead of the song comes the alarm-note, *Sharr-sharr.* Had you put your binocular upon him you would have found that the feathers on the top of the head are jet black, which can be raised as a crest; the nape of the neck ashy-grey; the back, wings, and tail a brownish-grey; throat and breast a light grey; under-parts white, as also are the under wing-coverts; legs a bluish lead-colour; the whole length of the bird between five and six inches. Female somewhat larger than the male, and the top of the head reddish-brown instead of black. The black-cap generally arrives in this country about the middle of April, and leaves us again in September.

In Guernsey the black-cap is commonly known as the Guernsey Nightingale.

THE WHEAT-EAR.

Another of the *Sylviidæ* often met with on moorland streams is the WHEAT-EAR or FALLOW-CHAT (*Saxicola œnanthe*), easily recognised as he flits away from stone to stone by the large white spot above the tail. The bird in the south often goes by the name of the Fallow-Chat, from its frequenting the upland fallows in search of food. It is a very early visitant, arriving in this country in the latter end of February or beginning of March, and stays till the end of September. On our southern downs the shepherds catch the birds by means of a horsehair snare placed under a sod of earth, formed into a hollow chamber, the bird having the habit on the least alarm—even a shadow from a passing cloud—of running into the first hiding-place it can find. Faber notices this in his "Ascent of Helvellyn:"—

THE WHEAT-EAR.

> "We watched one fair cloud sail
> For some Atlantic haven; the gay fir
> Looked through the mist below like gossamer.
> The wheat-ears ran or glided through the grass
> And o'er the stones."

THE WHEAT-EAR.

Formerly these birds were caught in great numbers on their migration, from the end of July—usually on St. James's Day, 20th July—to the middle of September, and esteemed as very great delicacies for the table. Pennant states in his time upwards of 1,840 dozens of wheat-ears were taken about the downs at Eastbourne, and sold for sixpence a dozen. Now they are much more scarce, and when exposed for sale at the poulterer's, fetch from 3s. to 4s. per dozen. John Taylor, the water-poet says :—

> "Th' are called wheat-ears—less than lark or sparrow ;
> Well roasted, in the mouth they taste like marrow.
> . . . .
> The name of wheat-ears, on them is ycleped
> Because they come when wheat is yearly reaped ;
> Six weeks or thereabouts, they are catched there,
> And are well nigh eleven months, God knows where?"

Gilbert White, generally so accurate, states that this bird remains with us during the winter; but this is not so. Its very early arrival in the spring probably was the cause of the mistake. The male bird, with the varied plumage of its blue-grey back, black wings, whitish buff of the under-parts, has a very handsome appearance, and when attending on its mate in the breeding season, utters a soft and pleasing song. The nest is generally placed in a heap of stones, or in a hole of a low wall, and sometimes in an old rabbit-furrow.

THE SPOTTED FLY-CATCHER.

The SPOTTED FLY-CATCHER (*Muscicapa—Musca*, a fly, and *carpo*, to take—*grisola*)—family, *Muscicapidæ*—arrives in this country about the third week in May; Selby says "when the oak-leaf is coming out," and may be constantly seen on the palings about woods, orchards, gardens, and lawns. It is called the Beam-Bird in Hertfordshire and other counties, Post-Bird in Kent, Bee-Bird in Norfolk, &c. The bird appears always intent on getting as many flies as possible, and it is a regular "dab" at it. Watch it for a short time; see how, with its sharp, hawk-like eye, it spots the various flies and beetles, and is down on the ground one moment and up in the air the next. Mr. St. John states that a pair of fly-catchers fed their young no less than five hundred and thirty-seven times in one day, beginning at twenty-five minutes before 4 A.M. and ending at ten minutes before 9 P.M. The call-note is harsh, like the word *Tshee*, several times repeated; it has no song. The colour of the bird is a uniform hair-brown on back, the head somewhat spotted with a lighter colour, the breast whitish and spotted, the lower parts greyish-white. In Somersetshire these birds are supposed to bring good luck :—

> "If you scare the fly-catcher away,
> No good luck will with you stay."

THE REDSTART.

We at times, more particularly when the stream flows by an old ruined abbey, or farm buildings, or old walls, come across a bird generally known as the REDSTART or

THE REDSTART.

FIRE-TAIL (*Ruticilla phœnicurus;* family, *Sylvidæ*). And we know him as he flies from us by his red tail, which is constantly in motion when it sits on an old wall or branch of a tree, watching your movements all the time, uttering its alarm-note. The redstart arrives in this country about the middle of April, and leaves again early in September. The male bird is very handsome, with his jet-black throat, dark blue-grey head and back and wings, white streak over the eye, bright, rusty red breast, rump, and tail, except the central feathers, white under-parts and dark legs. The female is a rusty brownish-grey, with the rump, upper tail-coverts, and tail, except the central feathers, orange-red—duller than in the male.

Seebohm says: "As the wheat-ear is the tenant of the cairns, the rocks, and the ruins of the wilds, in like manner the redstart may be designated a bird of the ruins and the rocks in the lower, warmer, and more cultivated districts. The redstart has rather a pleasing song, best heard in early morning, something like a wren's, only not so loud or buoyant, but its call-note, *Weet-tit-tit*, sharply uttered, is more familiar to us than its song."

THE HEDGE-SPARROW.

The HEDGE-SPARROW (*Accentor modularis*), Winter Fauvette, Dunnoch, Hedge-Warbler, is no sparrow, but one of the *Sylviadæ*.

This familiar bird, common everywhere, frequenting our hedgerows and gardens, remains with us all the year, and, like the robin, comes close to our habitation in the winter, but is not so bold. Its pleasant song may be heard at all times through the day and often far into the night. It is an early nester, and its compact nest with its blue eggs is the delight of schoolboys. It is the commonest of the foster-mothers of the cuckoo, that bird appearing to choose this nest for depositing its eggs more frequently than any other; and Drayton says that the hedge-sparrow is often devoured by its giant foster-child:—

> "The hedge-sparrow, this wicked bird, that bred,
> That him so long and diligently fed,
> By her kind tendrance, getting strength and power,
> His careful nurse doth cruelly devour."

And Shakespeare ("Henry IV., Part i. Act. v. sc. 1):—

"And being fed by us, you us'd us so
As that ungentle gull, the cuckoo-bird,
Useth the sparrow."

We can well recollect our first discovery of a hedge-sparrow's nest, and the delight it gave. As Wordsworth truly says:—

"Behold, within the leafy shade,
Those bright blue eggs together laid;
On me the chance-discovered sight
Gleamed like a vision of delight."

The bird has a very sombre plumage. The bill is dark

THE HEDGE-SPARROW.

brown; the head, nape and sides of the neck bluish-grey, streaked with brown; back and wings reddish-brown, overlaid with darker streaks of the same colour; eyes hazel; chin and throat bluish-grey; the breast and under-parts of a dusky white; legs a lightish transparent burnt-sienna colour. The female is a little more spotted than the male on the head. The bird feeds chiefly on insects, caterpillars, &c., in the summer months.

If our way lies by commons and heaths we come across a prettily marked bird, the WHINCHAT (*Saxicola rubetra*). It is one of our migrants, arriving in this country about the middle of April. You will recognise it by its call-note, resembling the word *U-tick*, many times repeated. It has rather a pleasant song, generally singing when sitting on the topmost twig of a bush or when hovering in the air.

The plumage is rather sombre, of a darkish-brown on head and back. The feathers have a light edging; a white streak, rather broad, over the eye; a dark-brown patch behind the eye and extending to the neck. The chin is white, the throat and breast of a delicate fawn-colour.

Seebohm says that the whinchat is one of the first birds to lose its powers of song. Singing incessantly through May, it loses its voice entirely by the end of June or the first days of July.

THE WHINCHAT.

CHAPTER VI.

THE BIRDS—(Continued).

THE GOLDFINCH.

AMONGST the family of the Finches which we meet by the river-side, none can compare both in beauty and in song to the GOLDFINCH (*Carduelis elegans*)—family, *Fringellidæ*. It is one of our most beautifully plumaged birds, its form and movements being extremely graceful, rendering it worthy of its specific name *elegans*.

This bird has many provincial designations, such as Goldspink, Thistle-Finch, Chalandire, Draw-Water, King Harry, Red-Cap, Proud Tailor, Fool's Coat, Sheriff's Man, Sweet-William; its Gaelic name is *Las-aer-Chrille* (Flame of the Wood), and from its sweet song and beautiful plumage is much sought after by the bird-catchers. Before the passing of the Wild Birds Act, as many as 1140 dozen, all cock-birds, have been caught in one season near Worthing. As a cage-bird it becomes very tame, and will learn many tricks:—

> "Live with me, love me, pretty goldfinch, do,
> Ay, pretty maid, and be a slave to you;
> Wear chains, fire squibs, draw water."

If you meet with one put your binocular upon him. Mark his scarlet forehead and throat, his satin-black crown and nape, with his broad white collar, his chestnut back, and black wings with broad yellow bars and white tips, his hazel eye, black tail with white outer feathers. It is a restless bird, flitting from thistle to thistle, uttering this call-note, *Twit-it, twit-it,* now clinging to the stem, now perched on the seed-vessel. Graham well describes him :—

> " But mark the pretty bird himself—how light
> And quiet his every motion, every note !
> How beautiful his plumes ! his red-ringed head,
> His breast of brown ! And see him stretch his wing,
> A fairy fan of golden spokes it seems."

THE GOLDFINCH.

Most of the goldfinches which are found in England migrate southwards in October, returning again in April.

THE CHAFFINCH.

The CHAFFINCH (*Fringilla cœlebs*) is another of the *Fringillidæ*, familiar to most of us, and constantly seen in the brakes and amongst the hedges bordering the river. In the "Countrie Farm," edition 1600, the chaffinch is called the *Spink*, and in the French work of Estreune and Liebault, the corresponding word is *Pinçon*. In Cotgrave's Dictionary, Pinson is a spink, chaffinch, sheldpate; and in More's "Suffolk Words," Spink is a chaffinch. In Hertfordshire it is called a Pink, or Twink, from its alarm-note

THE CHAFFINCH.

—*Pink-pink*. Nicolls, in his poem on the cuckoo, published 1607, says:—

"The speckled spinck that lives on gummie sap."

In Scotland the chaffinch is called the Shilfa. Graham, in his laudation of Mary Stuart's beauty, says:—

"Her cheek is like the shilfa's breast,
Her neck is like the swan's."

Put your glass upon him; look at the variegated plumage, the pinkish-red breast and dappled wing, with russet back, of the male bird.

The song is rather monotonous; but in Germany it is highly esteemed for its musical powers, particularly for its double trill, which is expressed by the words, "*Finkferlinkfink zischesia harvelalalalaziscutschia.*" In France, about Orleans, it is supposed to say, "*Je suis le fils d'un riche prieur;*" in Normandy, "*Qui est ce qui veut venir à Saint Symphorien;*" in Lorraine, "*Fi, Fi, les laboreux, j'virions ben sans eux;*" and about Paris, "*Oui, oui, oui, oui, je suis un bon citoyen*" (see Swainson's "The Folk-Lore of British Birds").

Three other members of this family are occasionally met with in our rambles. The LINNET (*Acanthus cannatina*), the REDPOLL (*Fringilla rufescens*), and the SISKIN or ABERDEVINE will flit across our path, or we may see them hard at work picking the seeds out of the plantains, or feasting on the down of the thistle and dandelion and other seeds.

The linnet generally confines itself to the waste places and commons. It is known as the Grey Linnet in some localities, the Brown Linnet in others—dependent on the season-change of plumage. In Scotland it is the Lintwhite, and in the mountainous parts of that country its place is taken by the TWITE, or MOUNTAIN LINNET (*A. flavirostus*), commonly known as the Hill-Lintie or Yellow-Neb Lintie.

The linnet has some sweet notes :—

> "But soft it trills amid the aerial throng,
> Smooth simple strains of sob'rest harmony;"

but its song has been much overrated.

The REDPOLL (*A. linaria*) is easily distinguished by its blood-red plumage on the crown of the head and its smaller size.

The restless little ABERDEVINE or SISKIN (*Chrysometris spinus*) is seldom seen, except in its migration south, but in the north of Scotland it is better known. It is much more green in colour, has a black head and parti-coloured wing.

THE HOUSE-SPARROW.

The HOUSE-SPARROW (*Passer domesticus*; family, *Fringillidæ*) is known to all in town or country. Impudent, yet wary; not easily caught or enticed, however much we may tempt with the most favourite food. "*Timeos Danaos nec dona ferentes*," they say, and yet we have seen in the Tuileries Gardens in Paris a man who, by whistling, brought all the sparrows round him. They perched on his hat, his shoulders, his arms, his wrists, and his hands; and even the cautious wood-pigeon came to his call. What innate mysterious power did that man possess?

W. Ralston, in his " Russian Folk-Lore," gives a legend about the sparrow:—" When the Jews were seeking for Christ in the Garden, all the birds, except the sparrow, tried to draw them away from His hiding-place; only the sparrow attracted them thither by his shrill chirruping; then the Lord cursed the sparrow, and forbade that men should eat its flesh."

The Bohemians have a set of charms to keep sparrows from the crops. We would advise the farmers to try them:—

1. Stick upright in a field a splinter cut from a piece of timber out of which a coffin has been made.

2. Lay a bone taken from a grave on the threshold or window-sill of your barn.

3 (and this really would not be difficult). If, while sowing, you put three grains of corn under your tongue, wait till you have reached the end of the furrow in silence, and then spit them out in the name of the d—l, no sparrow will come into your field, though your neighbour's may be full of them. (See Swainson's "Folk-Lore for British Birds.")

Skelton's poem on the death of Phyllyp Sparrowe, killed by his cat Gyp, written in 1508, bringing all the birds together to weep for his loss, proves how great a favourite the bird was at that time :—

> "The cat specyally
> That slew so cruelly
> My lytell prety sparowe
> That I brought up at Carowe."

THE YELLOW-HAMMER.

What pleasant thoughts of happy schoolboy days are brought to mind as the YELLOW-HAMMER (*Emberiza citrinella*)—family, *Emberizidæ*—flits before us from bush to bush, or utters its monotonous song from the top branch of a furze-bush, all golden with its blossoms! What visions of purple-streaked eggs and hair-lined nest, so artfully concealed at the foot of a bush or tussock of grass by the side of a bank!

HEAD OF THE YELLOW-HAMMER.

> "What tender memories are bound
> To this familiar hedgerow sound!
> The creature's homely glee
> Associates me with the hours
> When, so pure childhood willed, all showers
> Were sunshine showers to me."

Its well-known, monotonous song has been compared with the words—

> "A little bit of bread and no che-e-ese."

In Scotland it has another signification—

> "Whetel-te, whetel-te, whee!
> Harry my nest and the de'il tak ye."

It is called the Devil's Bird in Scotland, from a curious notion that it drinks three drops of the devil's blood every May morning, which is alluded to in the curious Scotch rhyme:—

> "Half a puddock, half a toad,
> Half a yellow yorling,
> Drinks a drap o' the de'il's bluid
> Every May morning."

The bird is called the Yellow Yorling in many parts of Scotland, and it is strange that the same superstition prevails in Bohemia. The bird is persecuted in that country because it always drinks some of the devil's blood on the 1st of May.

In Shropshire this bird is sometimes called the Writing-

THE WRY-NECK.

Master, from the irregular lines on the eggs, also the Writing Lark :—

> "Fine eggs, pen scribbled o'er with ink, their shells
> Resembling writing scrolls, which Fancy reads
> As Nature's poesy and pastoral spells."
> —*See Swainson's Folk-Lore.*

In France the bird is called *L'Ecrivain*.

THE WRY-NECK.

Towards the end of March or beginning of April one hears a curious sound amongst the trees difficult to describe.

THE WRY-NECK.

Dresser says it may be compared to the word *Hveed-hveed-hveed*, frequently repeated, sometimes loud, sometimes soft, but usually in a kind of prolonged plaint. Newton ("Yarrell," fourth edition) says: "The cry is not unlike that of the kestrel, and consisting of the notes, *Que-que-que*, many times and very rapidly repeated." This cry is the call-note of the WRY-NECK (*Junx torquilla*); family, *Picidæ*. The bird generally arrives a few days before the cuckow, hence the name of Cuckow's Mate given to it by country-folks. It is also known as the Snake-Bird, from the peculiar hissing noise it makes if disturbed when sitting on its nest, if nest it may be called; but the eggs are generally laid on the bare wood at the bottom of a hole in a tree; they are pure white and glossy. In some counties it is called the Emmet-Bird, from its love for ants as food. The name "Wry-Neck" arises from the peculiar habit of moving its head and stretching out or twisting the neck (as shown in the woodcut).

Although for the most part of one colour—a rich greyish-brown—it is so varied by the different shades as to form, as Bewick remarks, "a picture of exquisite neatness."

The bird is particularly fond of ants, and will settle down close to an ant's nest, and appears to be regardless of danger when thrusting its long tongue into the ant-heap. It will allow at this time a close approach, so that it could easily be killed by a blow, and we once, in early days, killed one of these birds whilst feeding as described with the end of our fishing-rod.

The bird has various provincial names, as Cuckow's Mate, Snake-Bird, Tongue-Bird, Dinnoch, Turkey-Bird. In France it is called *Torcol;* in Germany, *Natter Vogel*.

THE GREEN WOODPECKER.

The GREEN WOODPECKER (*Gecinus viridis*)—family, *Picidæ* —is occasionally met with as we pass to and fro to the river. We have often noticed these birds when fishing in Wiltshire and Hampshire, and when not too much disturbed will take the same round day after day, visiting the same trees, beginning at their base, and working round and round all up the trunk, searching for food. The bird is especially fond of ants, and hops in a curious upright position from one ant-hill to another, and when disturbed generally flies off, uttering its loud, laughing cry (once heard always to be remembered), to the nearest tree, there to creep up the trunk or on the branches, taking care to be out of harm's way. This laughing cry has given the bird the provincial name of the Yaffel:—

"The skylark in ecstasy sang from a cloud,
  And chanticleer crowed and the *yaffel* laughed loud."

THE GREEN WOODPECKER.

In Hertfordshire the bird is known as the Whetile or Cutter, which, according to Mr. Swainson, is derived from the Anglo-Saxon *thwitan*, "to cut;" but it also goes by the name of the Rain-Bird or Wetall, from the supposition that its peculiar cry is constantly repeated when wet weather is coming.

It has many other provincial names, as Cutbill, Woodspite, Awl-Bird, Woodwall or Woodweele :—

> "The woodweele sang, and would not cease,
>   Sitting upon the spray,
> Sae loude he wakened Robin Hood,
>   In the greenwoode where he lay."

Chaucer calls it the Woodwale—"With chalandre and with *woodwale*."

Nurdis, in "The Village Curate," says :—

> "Now we hear
> The golden woodpecker, who, like a fool,
> Laughs loud at nothing."

If you put your glass upon him you will find the male bird has a scarlet top to its head, black bill, black round

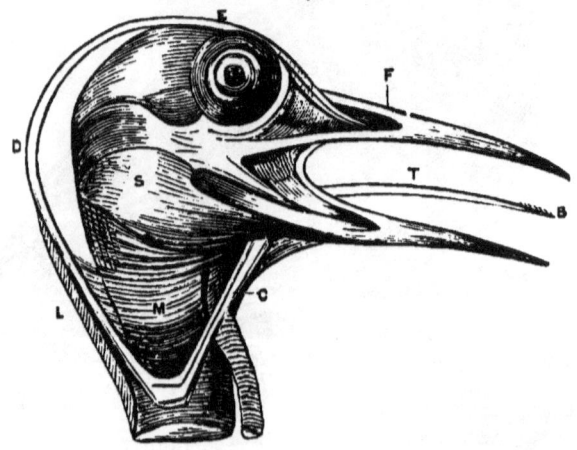

DIAGRAM TO SHOW THE TONGUE OF THE WOODPECKER.

T, the tongue.
B, the barbs of the tongue, pointing directly backwards.
C, D, E, F, slender osihyoides supporting the tongue turning into two long cartilaginous processes which form a very acute angle at their junction with the tongue ; bending downwards at C, they pass obliquely round the sides of the neck connected by a membrane M, then being inflected upwards, converge towards the back of the head, where they meet, and being enclosed in a common sheath at the cranium, E, till it arrives between the eyes. From this point the two cartilages, which are closely confined, are deflected towards the right side, and terminate at the edge of the aperture of the right nostril, F.

The cartilages are represented at D drawn out of the groove provided to receive them. The curvature is diminished by the muscles, L, and then the tongue is protruded.

the eye, and scarlet spot beneath; upper part of back dark glossy green; lower part and rump and upper coverts brilliant yellow; under-parts greenish-white; under-wing and tail-coverts whitish, irregularly barred across.

The eggs are placed at the bottom of a hole in the trunk of a tree, which the birds bore out to the extent of sometimes a foot. They are of a pure white. At times these birds have to fight with the starlings, who wait till the hole is completed and then take possession, and are often successful in keeping it.

The structure and position of the tongue of the woodpecker is peculiar. It is capable of being protruded to a great length, and this is effected in the manner shown in the figure on page 126.

THE GREAT SPOTTED WOODPECKER.

The GREAT SPOTTED WOODPECKER or WOODPIE (*Picus major*), now called *Dendrocopus major*, is not so common, but occasionally his tap-tap against a decayed branch or stem of oak or ash may be heard. It is sometimes seen searching the alders and pollar willows by the sides of the streams for its insect food, for although at times it will take fruit, especially cherries, yet insects form its principal diet. In spring-time this bird makes a very peculiar jarring noise, at one time supposed to be caused by placing its beak in the cleft of a branch, but it is now ascertained to be the result of a rapid repetition of taps with its beak, and is evidently a call-note, and sounds like, *Er-r-r-r-r-r* often repeated, quite different to its usual note, which, according to Yarrell, sounds like *Quet-quet* or *Gick-gick;* according to Dresser, *Tick* or *Tchick*.

The bird is very quiet in all its movements, and the rapid way it ascends, always spirally, the bole of a large oak up to its topmost branches is surprising, and then, as it were, suddenly falls, and just with a rapid movement of its wing, directs itself to another tree, and the same plan is repeated.

The male is a very handsome bird. The irides are bright red; the forehead buff; ear-coverts and round the eyes white; top of head jet black, with a bright scarlet patch on the occiput; plumage on back black and white; middle of

THE GREAT SPOTTED WOODPECKER.

throat, breast, and under-parts dirty white; legs, toes, and claws greenish-grey. The female has no red on the head.

The LESSER-SPOTTED WOODPECKER (*D. minor*) is more common than its larger cousin, and if looked for carefully amongst the elms and poplars, may often be noticed. It frequents the valley of the Thames and many of our southern stream-valleys. It is more barred on the wing, with the back more white than the Greater; the crown of the head red.

CHAPTER VII.

THE family *Paridæ*, the Titmice, are fond of the waterside, and we are pretty sure to see some of them and hear their call-note as they flit from tree to tree seeking their food.

THE MARSH TITMOUSE.

The MARSH TITMOUSE (*Parus palustris*) is most frequently met with among the low alders and bushes on

THE MARSH TITMOUSE.

the river-side. The bird is easily distinguished from the coal titmouse from having the crown of the head entirely black, the back and wing-coverts brown, the throat and breast a dull white.

This bird may also be easily recognised by its call-note, as Alston says, a long-drawn *Pey-pey*. Seebohm gives it as *Tay-tay-tay-tay* in rapid succession. This author also says: "The marsh tit has scarcely a right to its name. It is never seen in the reeds or in the sedge, which are the special characteristics of the marsh, but in bushes and trees of all kinds, great or small, on the confines of the reeds, on the bushes by the river-side, or in the garden. Even in the suburban gardens of London or Sheffield it is almost sure to be found. Nevertheless, it is less partial to very dry districts than some of the other tits."

Gould says that it is common in the neighbourhood of the Thames and other rivers.

THE COAL TITMOUSE.

The COAL or COLE TIT (*Parus ater*) is a very lively bird, flitting from bush to bush, uttering its call-note, *If-hee, if-hee, if-hee,* creeping up the stems examining every crannie and crack in the bark, first head upwards, then head downwards, then hanging by its leg at the end of a branch, always at work, always busy; and having thoroughly investigated every likely spot, off he flits to the next tree, to go through the same process. Put your glass upon him, and you will find the coal-tit has a black head, the nape of the neck with the ear-coverts white, the upper parts brown, the wings and tail a greyish-brown; there is a double bar of white across the wings; breast and lower part white, brownish-yellow on the sides. The coal-tit is distinguished from the marsh by the *white patch on the nape of the neck* and by its different call-note. He is a little bit of a bird, only about four inches long, including his tail.

As regards its British name, Coal or Cole Tit, the editor of the fourth edition of "Yarrell's British Birds" has the following note:—"Merritt in 1667 called this bird, Latinising its name, *Carbonarius.* The French, *Charbonnière,* applies to this as well as to the great titmouse, and equally shows the meaning of the word, which most later authors have spelt 'cole;' but as it has clearly nothing to do with cole, the plant (as found in cole-wort and coleseed), and we have given up spelling the name of the fuel we burn otherwise than 'coal,' it is wrong to keep 'cole' as the distinguishing prefix of this titmouse. It may be urged that the Germans set us the example, writing *Kohl-Meise* and not *Kohle-Meise*, but here the *e* is doubtless dropped by way of abbreviation or euphony. It may also be remarked that the second syllable of the word tit-mouse has nothing to do with the quadruped so-called, but it is cognate with the root of the French *Mésange*, the Anglo-Saxon *Mase*, the German and Danish *Meise*, the Swedish *Mes*, and the Dutch *Mees* (*pl.*

THE COAL TITMOUSE.

Meesen). It may, therefore, be doubtful whether the plural of titmouse should be titmice, as custom has it; but the editor has not the courage to use 'titmouses,' though he believes he has heard East Anglians say 'titmousen,' just as they always use the old form *housen* for *houses!*"

In Scotland the bird is called Coaly-Hood.

Seebohm gives the scientific name as *Parus Britannicus*, and says it differs from the typical *Parus ater*, which appears continually to visit our islands, and apparently to inter-

132 THE RIVER-SIDE NATURALIST.

breed with the British sub-species, in having the slate-grey of the upper parts suffused with brown, which in the typical form is observable on the rump only, and by having more brown on the flanks.

The food of the coal-tit is chiefly composed of insects, but Mr. Tegetmeier says that it will eat seeds, and gives an instance of one feeding on filberts.

THE LONG-TAILED TITMOUSE.

We at times come across a family of the LONG-TAILED TITMOUSE (*Acredula rosea*) as they search for their food in

THE LONG-TAILED TITMOUSE.

the trees and hedgerows. This bird has many provincial names: Bumbarrell, Bottle-Tit, Feather-Poke, Long-Tailed

Mag, Huckmuck, Mum-Ruffin, &c. It builds a beautiful domed nest, and lays a great many tiny eggs—sixteen to seventeen. It is a very restless, active bird; the brood keep together for some time, and as they fly from bush to bush keep up a continual twitter, a kind of plaintive but rather shrill note. Its long tail at once distinguishes it from the other *Paridæ*. It is very prettily marked. Eyelids are bright orange-red; front and crown of the head dull white, streaked with black; upper part of back black; lower part a dull rose-red; breast dull white, with some black marks; under-parts a beautiful roseate tint; legs black; the middle feathers of the tail black; outer white. Dresser says that the *Acredula caudata*, for which it has been sometimes mistaken, is extremely rare in this country.

THE GREAT TITMOUSE.

The GREAT TITMOUSE (*Parus major*), also known as the Ox-Eye, Saw-Sharper, &c., is rarely seen near the rivers or in the open country, inhabiting chiefly woods and gardens and sheltered enclosed districts. The bird is common enough, is an early breeder, and its peculiar chirpy notes are often heard early in February. Seebohm says: "If you wander out in the fields and woods

THE OX-EYE OR GREAT TITMOUSE.

of a winter's morning, the sharp, unmistakable note of the 'Ox-Eye' will most probably be the first sign of bird-life you notice."

The great tit is very fond of fat, and can be often tempted near the house by hanging out a bone or a piece of suet or bacon. Others of the *Paridæ* are equally as fond of these morsels, and will also be tempted; but the great tit will lord it over all the other birds present, except the robin, who fears nothing of his own size.

THE BLUE TITMOUSE.

The BLUE TITMOUSE (*Parus cœruleus*), Blue-Cap, Billy-Biter, is one of the prettiest of our *Paridæ*, and a very common species. All of us who, in our boy days, have been fond of bird-nesting can well remember with what terror we first heard that peculiar hissing when inspecting

THE BLUE TITMOUSE.

a hole in a tree or wall, firmly believing that a snake and not a bird was the cause of it. When better informed, we knew it was Mistress Blue-Cap, who defends her home with great courage and pertinacity. The nest is at times placed in very curious localities; an old bottle, even a country post-office box, has been appropriated. The crown of the head is bright cobalt blue; the forehead white, as well as a line of feathers extending round the blue crown; cheeks and hinder part of the neck white; back green; wings and tail blue; breast and under-parts sulphur-yellow; legs bluish-grey. Wordsworth appreciated its bright colours:—

> "Where is he, that giddy sprite,
> Blue-cap, with his colours bright,
> Who was blest as bird could be,
> Feeding on the apple-tree;
> Made such wanton sport and rout,
> Turning blossoms inside out;
> Hung head pointing towards the ground,
> Fluttered, perched, into a round
> Bound himself, and then unbound;
> Lithest, gaudiest harlequin,
> Prettiest tumbler ever seen!
> Light of heart and light of limb,
> What is now become of him?"

CHAPTER VIII.

THE ROOK.

THERE are but few, if indeed there are any, of our rivers of any size without one or more rookeries in the trees which grow so luxuriantly in the valleys through which they run.

The ROOK (*Corvus frugilegus;* family, *Corvidæ*) never fails, therefore, to put in an appearance. His glossy black plumage attracts our attention as he struts over the water-meadows seeking for slugs and worms, larvæ and the like.

The rook, or, as he is more commonly called, the Crow or Craw, was in former days the most maligned of the feathered race; he sucked eggs, ate the young chicks in the poultry-yard, devoured the new-sown corn, and had the audacity to follow the husbandman in the field and try and deceive him by his friendship; he tore up the roots of the fresh-sprouting wheat out of mischief, and gobbled up all the fruit on the trees. In fact, at one time he was considered such a delinquent that every man's hand was against him. In the reign of bluff King Harry rooks and crows were so numerous, and were thought to be so detrimental to the farmers, that an Act was passed for their destruction. Every hamlet was to provide crow-nets for two years, and the inhabitants were obliged at certain times to assemble and concert measures for their extermination.

Ray, in his edition of "Willughby," says: "These birds are noisome to corn and grain. If rooks infest your corn, they are more terrified by taking a rook and plucking it limb from limb in their sight, and then casting the several limbs about your fields, than if you hung up half a dozen dead ones."

As in the old days, so again at this present time, want

THE ROOK.

of knowledge of their food and habits is causing the wholesale destruction of these birds in many parts of England and Scotland. This is a great mistake, for whatever mischief they may do to the cereal crops in seasons when they are pressed for food, this occurs but seldom. There can be not the slightest doubt that they destroy an enormous number of the larvæ of the wireworm, of slugs, beetles, and many other much more potent enemies to the various agricultural products than the rooks.

A correspondence appeared in the *Field* newspaper some short time since in relation to the rook's capability of conferring benefit or the reverse on the agriculturist; it is of so much interest that we have taken the liberty of reproducing it. Mr. Speedy writes:—

"Whether rooks are the foes or friends of the farmer has long been a controverted question. Some assert they do a very considerable amount of damage to crops, while others maintain that any mischief they do is more than counterbalanced by the immense numbers of slugs, wireworms, &c., they devour, which are so destructive to young plants. That they do a certain amount of mischief, especially in dry seasons, is true; but they are often mistakenly charged with eating up the crops while they are taking the best possible means of protecting them. In illustration of this, some years ago, in the early spring, a farm-grieve tried to stalk a flock of rooks which were busy feeding on a field of grain. Meeting him, he requested me to shoot some of them, as they were 'playing the vera mischief wi' the wheat.' Concealing myself behind a hedge, I asked him to go round and startle them, to which he readily agreed, when I had no difficulty in killing a couple as they flew overhead. 'Man, that's grand!' the grieve exclaimed as he came forward; 'I'll hing them up in the field to scare ithers.' Carrying the two birds in my hand, we walked to where the rooks had been feeding, when we discovered numerous holes dug by their powerful beaks, and blades of young wheat strewn all around. On minute examination, however, I observed that the blades which had been pulled up corresponded with a considerable number which were not so bright in colour as healthy plants are, and digging one up with my knife, I discovered a small grub adhering to the root. On opening the gizzard

THE ROOK.

of some of these birds I found a large number of grubs, while no traces of wheat or green blades were discernible.

"Another illustration in point. During the protracted drought of last summer a blight seemed to come over a large quantity of onions in a market-garden at Craigmillar. The plants at first became slightly discoloured, and eventually withered away. My attention was called to the circumstance that the market-gardener was having his onions pulled up and eaten by the rooks. I felt that, while it might be true that the onions were being pulled up, I was, at the same time, certain that it was not that they might be partaken of as food. On visiting the spot I at once observed that the onions were blighted and fading away, as if they had been sown where there was neither moisture nor soil. This at once led to the solution of the difficulty, as, upon careful inspection, I found, as I had anticipated, that the onions were being destroyed by grubs, which, in incalculable numbers, pervaded the entire area on which they had been sown. It will thus be seen that the object of the rooks in pulling up the plants was to devour these pestilent insects, as not one of the plants exhibited the slightest indication of having been partaken of. It was interesting to note the sagacity which the birds displayed in pulling them up, as in no case did they make a mistake, even where discoloration could not be discovered by the human eye. It is needless to say that here, as in the case of the wheat referred to, the rooks rendered a valuable service to the farmer and to society at large. It will be evident that there is a danger of the interests of the farmers being overlooked by superficial observers rushing to hasty conclusions, as was the case of the grieve referred to. As by scientific investigation many palpable mistakes in agriculture are being discovered, so will the interests of farmers and gardeners be promoted as the facts of natural history become more generally and accurately understood."

Mr. M'Bean, although evidently a lover of rooks, has his doubts, and answers Mr. Speedy as follows:—

"Whether the rook is the friend or the foe of the

husbandman has, according to Mr. Speedy, in your issue of February 25 last, 'long been an open question,' but whether now finally set at rest I am not aware. Naturalists—and I suppose I may include Mr. Speedy among the number—differ among themselves on many subjects in natural history, which shows that they are not infallible; but I believe I am correct in stating that, so far as the rooks are concerned, they are about unanimous in the opinion that these birds are the friends and not the enemies of the farmers. Mr. Speedy, however, does not furnish data sufficient to enable one to decide one way or the other. He admits that 'rooks were responsible at times for a very considerable amount of mischief;' but, on the other hand, he states that these birds 'were charged with eating up crops while taking the best possible means to protect them.'

"As an old farmer, I should like to be informed what crop it is that the rooks are charged with eating up while taking the best possible means for its protection. I am not aware of such generous action on the part of the rooks. It is, in my opinion, neither wheat, barley, oats, potatoes, nor turnips. Neither is it the crop of onions alluded to by Mr. Speedy, for, be it noted, this crop was 'blighted and fading away as if it had been sown where there was neither moisture nor soil'—that is to say, this crop was already destroyed before the rooks attacked the grubs with which the crop was infested. The wily birds were wide enough awake to the fact that it was no use searching for grubs so long as the crop presented a healthy appearance, for at this period the grubs were either absent, or else so diminutive as to be entirely beneath the notice of the big birds. It is thus clearly seen that the rooks made no effort to protect the onions, but assisted the grubs in destroying the crop. As to the field of wheat alluded to, Mr. Speedy does not seem to have realised the fact that in pulling up the young braird the rooks were in search, not of grubs, as a few suppose, but in search of the parent seed, and this pulling up and digging continues so long and no longer as any substance remains in the grain, which no sooner is exhausted than the rooks forsake the field till harvest approaches.

"At one period I had ample opportunity of determining whether the rooks are friends or enemies, and after careful investigation arrived at the conclusion that the produce of the farm, when obtainable, forms their staple food, and for such favours they do little or nothing in return for their keep ; and further, that one-half, or, more correctly, two-thirds of the animal food consumed by these birds consists of the friends and not the enemies of the crops. It were well, therefore, that those who consider these bird friends defend and demonstrate more clearly than they have hitherto done the character of this, the most interesting British bird we have. No one could be more pleased if this can be accomplished than the writer, for these birds are prime favourites."

Another writer says : "While, as I have here pointed out, they render immense service to agriculturists in picking up wireworms and grubs, which are so destructive to plants, it is nevertheless true that in certain seasons they are responsible for a very considerable amount of mischief. When potatoes are appearing through the ground, they dig down for the end, which they rarely fail to carry off, and in consequence numerous blanks are visible when the crop grows up, unless vigilance is practised by 'herding' them. In the plundering of potato-fields rooks display a more than ordinary degree of sagacity in their mode of getting at the early potatoes. Instead of digging down along the side of the plant from the top of the ridge, they are often to be found penetrating into the sides of the ridges at a lower level, right opposite the potatoes, so that labour is thereby economised.

"In protracted droughts, as in hard frosts, rooks have extreme difficulty in obtaining their food-supplies. This I frequently noticed by about a score of them coming regularly to feed in my back-garden, where scraps were thrown out to the birds. No sooner, however, had there been a few hours' rain than they disappeared, preferring grubs and worms—the catching of which was facilitated by the moisture—to the bits of bread and meat thrown out. So long as the weather remained damp, with occasional showers, they were never seen; but in dry weather, as in frost, they immediately returned. The reason of this is obvious, as in dry, hot weather grubs go down into the cool earth beneath, but invariably return near the surface after rain.

"It is asserted by some agriculturists that the damage done to young wheat by rooks is not by eating the seed, but by nipping off and devouring the shoot, which, of course, destroys the plant. If such were the case, it is not too much to say that when pressed by hunger in droughts or frost they would regale themselves on the shoots of grain or grass, which they would have no difficulty in obtaining. This assertion I am exceedingly loth to believe, as in none of those whose gizzards I have examined have I ever found green blades of any description."

The intelligence of rooks is shown in many ways. They appear to be capable of distinguishing a gun from a stick when carried in the hand, at once alarmed by the former, but perfectly indifferent to the stick.

Watching rooks being fed in a garden, a writer observes: "It is in such circumstances that their shrewdness and forethought obtrude themselves on our attention. Several cats were in the habit of appearing as soon as the food was thrown out. The rooks, by their noise and attitude of offence, proved themselves able to keep the cats at bay until their appetites were fully satisfied, when, on their leaving, the cats picked up the remaining food. On the rooks discovering this, it was an interesting study to observe them as, after having satisfied themselves, they picked up pieces of the remaining meat and carried them off to different parts of the garden, where they carefully buried them in the earth. During the afternoon they regularly returned, and with unerring accuracy disinterred the pieces of meat buried in the morning. The reflective instinct exhibited by the rook is proverbial, and developed to a larger extent than in most other birds. Its powers of arithmetical calculation have long attracted the attention of naturalists. It has been found that they can count numbers accurately up to three inclusive, but that this is the limit of their capacity of calculation. This peculiarity in the rook has been discredited by many, but, when tested by experiment, has been again and again verified. For example, when they are so pressed for food during a snowstorm as to visit a stack of grain, let a place of concealment be extemporised by branches of trees, or other material, within easy shot, where watchers can successfully conceal themselves. If, after being repeatedly fired at from the ambush in question, the rooks discover one, two, or three persons betake themselves to the place of concealment and leave at intervals, it will be found that they will not descend to feed until the last of the three has left. But should four or more persons place themselves under cover, it will be found that after the third has left their sense of danger disappears, as will be seen by their beginning to feed with apparent security. I am not aware of this peculiarity being possessed to the same extent by any other bird."

Another instance of their intelligence was related to us by the observer: "After the hay is carted off a fresh-mown meadow, the rooks assemble in considerable numbers and dig with their beaks small holes all over the field. By the next morning every one of these holes contains a small white slug (*Limax agrestis*), and as soon as daylight appears down come the rooks, and without any trouble procure their morning meal, very much to their own and the farmer's benefit."

Rooks feed chiefly on the larvæ of numerous insects, worms, slugs, snails. They are particularly fond of the larvæ of the cockchafer, crickets, wireworms, daddy longlegs, &c. They are also partial to some kinds of fruit, such as cherries and walnuts before the shell hardens. In Scotland they take the crowberry. In hard winters and in very dry springs the rook becomes omnivorous, and will then take offal of all kinds, and follow the hog-pail or suck an egg, but that in the spring he will search the fields and hedgerows for pheasants' or partridges' eggs requires much better evidence than the assertions of interested gamekeepers or their masters, who probably do not know the difference between the rook and the carrion crow or jackdaw, both of these latter being habitual pilferers and devourers of eggs and young birds.

The late Lord Erskine was so convinced that this bird was the farmer's friend, that he wrote a poem on the subject, in which he makes the rook address him, after the bailiff had been dealing destruction on a colony of these birds :—

> "Touch'd with the sharp but just appeal,
> Well-turn'd, at least, to make me feel,
> Instant this solemn oath I took—
> No hand shall rise against a rook."

Rooks are very early breeders, and begin repairing their nests at the first change from wintry weather. Gilbert White alludes to this in his verses on the dry warm weather in winter :—

> "Sooth'd by the genial warmth, the cawing rook
> Anticipates the spring, selects her mate,
> Haunts her tall nest trees, and with sedulous care
> Repairs her wicker eyrie—tempest-torn."

This building-time is one of constant flurry and excitement, of battles and of thefts. The propensity to thieving is at this time made remarkably evident. The old birds are well aware of this, and never leave the nest unguarded. But mark what happens to a pair of young birds entering for the first time on their domestic duties. Too confident in the honesty of, perhaps, their own parents, building close by them, they both leave the nest to procure materials; and when they return, where is their loved dwelling? All has disappeared; the wily old birds have robbed them of every stick.

This custom of the old birds was noticed by Ray, who, writing in 1776, says: "I have been told by a worthy gentleman of Sussex, who himself observed it, that when rooks build, one of the pair always remains to watch the nest, else if both go, their fellow-rooks ere they return will have robbed and carried away all the sticks and whatever else they had put together;" and he pertinently adds:

"Hence, perhaps the word rookery with us is used for cheating and abusing."

What labour and how many flights this nest-building causes! How wonderful the twigs are interlaced, and the bottom and lining made heavy and warm, to prevent destruction by the wind, and to keep up heat during incubation on the many cold wintry days of early spring!

The following is the result of a careful examination of a rook's nest of the year, taken from the rookery at Cowdray Park, where the trees are mostly beech:—

Size of Nest.

Breadth outside	15 inches.
„ inside	7 „
Depth outside	$5\frac{1}{2}$ „
„ inside	$4\frac{1}{2}$ „

The inner part or hollow of the nest was entirely composed of the smaller twigs of the beech, interlaced in every direction, and on these was the lining, consisting of dried leaves and decayed moss, mixed with a little mould, as is found at the base of the trees amongst the external roots; also a small quantity of decayed grass, the whole mass being about $1\frac{1}{2}$ inch thick, and weighing about 10 ounces.

The number of sticks and twigs were 493; their weight, 2 lbs. 1 ounce. One hundred and fifty of these sticks were from 11 to 23 inches in length, the thickest twig (oak) 15 inches long and 1 inch in circumference. The twigs, both oak and beech, were all fresh wood of the year.

Rooks at times desert their nests and the trees they have been accustomed to. The cause is often a mystery. A remarkable instance occurred in the spring of 1889:—

"The Elm rookery at Stapelgrove has been completely deserted by the rooks—to the number of one hundred and fifty—apparently in one day. This might well be considered a very ominous event, presaging death and disaster to the proprietor; but happily a different explanation can be given. Some men working near observed a pair of crows pillaging the rookery, driving away the rightful owners, and sucking their eggs. The working-men did not fully comprehend the meaning of the attack, and omitted to go to the rescue in time. Every nest now has its empty egg-shells, and not a rook remaining. Naturalists may like to know that the date of the burglary was the 5th April."

It is very interesting to watch the flight of these birds, particularly in the evening, when they return to their rookeries after the day's work, streaming home in twos and threes. Those already assembled receive each lot with much fuss and noise, and when the greater part are collected together, of a sudden up they all rise and gambol in mid-air before retiring to rest.

> "Behold the rooks—how odd their flight!—
> They imitate the sliding kite,
> And seem precipitate to fall,
> As if they felt the piercing ball."

This propensity to shoot down with closed wings is supposed to foretell rain. A French proverb says:—

> "Quand le corbeau passe bas,
> Tous l'aile il porte la glace;
> Quand il passe haut,
> Il porte le chaleur."

Black as he looks, his plumage is richly glossed with purple on the upper parts, particularly on the head and

FIG. 1.—HEAD OF THE ROOK.   FIG. 2.—HEAD OF THE CARRION CROW.

neck; the beak is black, and in the adult bird the forehead, lores, chin, and throat are bare, the skin being scabious and of a grey hue. In the carrion crow the feathers cover all roese parts, and thus can be distinguished; but young thoks are also feathered at the beak until the first moult.

THE JACKDAW.

The JACKDAW (*Corvus monedula*), another of the family, is constantly associated with the rook, building in the holes of the trees in which the rooks have their nests, as well as in church-towers, ruins, and cliffs, seeking the same kind of food; one seldom sees a flight of rooks without the lesser bird being amongst them. The jackdaw is a sad pilferer, and particularly fond of other birds' eggs and their young; he has been seen to take the callow young from a pigeon's nest, break open their skulls, and devour the brains,

THE JACKDAW.

leaving the rest of the body untouched. We have seen him search the Corinthian columns of the houses in some of the London squares, and bring out and eat young sparrows one after the other, to the great distress of the old birds, who dare not attack him. According to Ovid, the jackdaw was the bird of Minerva, but was displaced in favour of the owl, owing to the jackdaw having told tales; hence he is designated "chattering."

The bird is much smaller than the rook, and is at once distinguished by his grey head and neck. He is very fond of perching on the vanes of the steeples of churches:—

> "A great frequenter of the church,
> Where, bishop-like, he finds a perch
> And dormitory too."

There are two members of this family which, owing to the present rage for high-game preserving, are now but seldom seen, as their well-known propensities for eggs of all kinds have brought them to the verge of total extinction—

THE MAGPIE AND JAY.

The MAGPIE (*Pica rustica*) and the JAY (*Garrulus glandarius*). The former's chatter and the latter's scream are now almost things of the past.

The magpie, or magot, as it is often called, was always considered a bird of ill omen, this prejudice prevailing in all countries in which it is found. It was the only bird that would not enter the ark, but sat outside chattering over the drowned corpses. To see one magpie at a time is supposed to be most unlucky, and the only way to avert any misfortune is to stop and take off your hat, making at the same time a profound bow. Sir Humphry Davy, in "Salmonia," says: "It is always unlucky for anglers to see a single magpie in the spring, because it indicates showery, cold weather; but when two appear the weather is mild and warm, and more favourable to fishing." It is a pity that so handsome a bird, and so lively withal, should be so persecuted.

The jay is more common, probably from frequenting the innermost parts of the woods and being extremely wary, but it has much diminished in open places and by the water-side. It was supposed to be partial to acorns, hence its name *Glandarius*. It is a beautiful bird, with a very unpleasant voice, addicted to sucking the eggs and eating the callow young of other birds, and therefore no mercy is shown.

THE WOOD-PIGEON.

There is no more pleasant sound than the deep *Coo-roo, coo-roo* of the Ring-Dove or Wood-Pigeon (*Columba palumbus;* family, *Columbidæ*). Campbell call its note "the deep, mellow crush." Its name, ring-dove, is derived from the beautiful iridescent patch, almost a ring, on both sides of the neck. It is also known in some counties as the Cushat, Queest, and Culver. Burns says:—

HEAD OF THE WOOD-PIGEON.

> "Thro' lofty groves the cushat roves;"

and Queest is probably from the Latin, *questus*, a complaining, the note being considered by some as melancholic:—

> "On lofty aiks the cushats wail,
> And echo coos the dooful tale."

Culver is a much older term, as we find Edmund Spenser using this word:—

> "Like as a culver on the bared bough
> Sits mourning for the absence of her mate."

It is worth while, when you hear his *Coo-roo, coo-roo*, thrice repeated, and ending with a short *coo*, to stop a moment. If he is answered you will see him, with a loud flap-flap of his wings, soar up to a considerable height and then float away, descending to where his mate is waiting for him, and this is often twice repeated; and then, just as he appears about to perch, up he goes again, and flies to some more distant tree.

What a fine bird he is!—bold and wary; difficult to see when perched amongst the trees, and only makes his presence known by his loud flight as you come upon him.

This bird has lately become very common in the London parks, as many as forty having been seen together on the lawn by Rotten Row in Hyde Park, apparently taking no heed of the "madding crowd."

THE DOTTEREL.

The DOTTEREL (*Charadrius morinellus;* family, *Charidriadæ*)—(the two Latin words literally mean "the dull bird, which, if looked upon, cures one of the jaundice." *Charadrius* is so translated in Ainsworth's Latin Dictionary)—is occasionally met with on the Cumberland and Westmoreland moors.

Whether dull or not, the bird furnishes the artificial-fly makers with feathers, taken from the wings, the colour of which they can get from no other source, and, therefore, a fly made of a dotterel-wing is highly prized. Bewick

THE DOTTEREL.

says: "The dotterel is said to be very stupid, and easily taken with the most simple artifice, and that it was formerly the custom to decoy them into a net by stretching out a leg or an arm, which caught the attention of the birds, so that they returned it by a similar motion of a leg or a wing, and were not aware till the net dropped over them, and covered the whole covey." This is taken from Willughby, who translated it from Gesner's *Historiæ Animalium*. Willughby says: "It is taken in the night-time by the light of a candle by imitating the gesture of the fowler. For if he stretches out an arm, that also stretches out a wing; if he a foot, that likewise a foot; in brief, whatever the fowler doth, the same doth the bird." Mr. Heysham, in his account of this bird, says: "They permitted us to approach within a short distance without showing any signs of alarm; in short, they appeared so indifferent with regard to our presence that at last my assistant could not avoid exclaiming, 'What stupid birds these are!'"

The old poet Skelton called the dotterel "that folyshe pek," and Drayton, "the sottish dottrell, ignorant and dull." Drayton also knew how good the bird is for the table: "The dottrell, which we thinke a very dainty dish."

In Wiltshire, when dotterels were not so scarce, the bird's movements were considered prognostic of change of weather :—

"When dotterel do first appear,
It shows that frost is very near;
But when the dotterel do go,
Then we may look for heavy snow."

In the early part of the sixteenth century there is a notice in the "Northumberland Household Book:"— "Item: Dottrels to be bought for my Lorde when they are in season, and to be had at 1d. a-pece."

THE CURLEW.

On the wilds of Exmoor and Dartmoor, and on most of the moors of the northern counties in England and Scotland, as well as in Ireland, the CURLEW (*Numenius arquata*), one of the *Scolopax* family, may be found breeding, and when disturbed utters its oft-repeated plaintive cry, as if you whistled the word *cur-lieu*.

They are very wary birds, and difficult to get at, and their warning whistle sets all other birds in the neighbourhood on the *qui-vive*.

You will know the curlew "by the length of his bill," as well as by his light-grey spotted plumage, large dark-brown eyes, and peculiar cry. The breast is much lighter than the rest of the plumage, and more spotted, the underparts as well as just above the tail being almost white. The female is somewhat larger than the male.

Buckland, in his "Curiosities of Natural History," says: "The sad, wailing cry of these birds while on the wing in the dark, still nights of winter, resembling the moans of wandering spirits, is believed in some parts of

THE CURLEW.

England to be a death-warning, and called the cry of the seven whistlers:"—

> "The curlew screamed above:
> She heard the scream with a sickening heart,
> Much boding of her love."

In Scotland the bird goes by the name of the Whaup, and is considered very uncanny. Sir Walter Scott, in "The Black Dwarf," makes Hobbie Elliott say, "What needs I care for Mucklestane Muir ony mair than ye do yoursel', Earnscliff? To be sure they say there's a sort of worricows and lang-nebbit things about the land; but what need I care for them?"—the lang-nebbit things being the curlews, supposed to transform themselves into goblins which go about houses after nightfall.

The Scots also had a notion that the young curlew when first hatched ran about with part of the shell still adhering to their heads. In "The Abbot" Adam Woodcock is made to say, "I believe in my soul you would run with a piece of the egg-shell on your head like the curlews, which we used to call whaups."

The young curlews, before they go to the sea, are excellent eating. In Lincolnshire there is an old rhyme:—

"A curlew lean or a curlew fat
Carries twelve pence on her back."

Of the *Falconidæ* we do not meet with many species in the southern districts, for although occasionally the HOBBY (*Falco subbuteo*) or the MERLIN (*Falco æsalon*) may cross our path, yet those commonly met with are the SPARROW-HAWK and the KESTREL.

THE SPARROW-HAWK.

The SPARROW-HAWK (*Accipiter nisus*), the dread of the farmyard, sometimes wings its rapid flight close to the bushes by the river-side, ready to pounce unawares on any of its denizens. It is the male, or smaller bird, we generally see. The female, much larger than the male, confines herself to the farmyard or game-breeding paddocks and pigeon-houses. She remains near the inhabited districts, and as long as a chicken or a young pheasant is left will never desert the locality, unless compelled by wounds or death. It

HEAD OF THE SPARROW-HAWK.

is her body which is the constant companion of cats, stoats, &c., on the gamekeeper's gallows-tree.

The male, very much smaller, keeps to the hedgerows and copses. His appearance at once sets the whole of the community of small birds into fearful alarm. He is the terror of the sparrows and finches, the dire enemy of all the swift-winged birds, and the swallows seem to take especial delight in mobbing him whenever they can, their scream of alarm giving warning to the general confraternity; but by his stealthy and quiet manner of approaching his prey he is pretty sure of his game. Knox ("Ornithological Rambles") says: "The depredations of this little tyrant of the woods and groves certainly surpass those of any other British bird of prey." Tennyson calls him "the hedgerow thief."

Seebohm says that "birds do not form the sparrow-hawk's only fare; sometimes you see him dip silently and swiftly down amongst the marshy vegetation of old watercourses and bear off a rat or a frog." Young rabbits and leverets fall to his unerring swoop, and in Scotland he fearlessly attacks the wood-pigeons, and does some good in this respect to the farmer; and by taking the weakly game-birds it helps to keep disease away, and preserve that healthy standard of perfection which nature inexorably demands. But, on the other hand, no rapacious bird is more to be dreaded by the gamekeeper or the chicken-breeder.

The sparrow-hawk has a mythological history. Nisus was transformed into this bird after his daughter's treacherous conduct, she being at the same time changed into a lark, so that the two should be continually antagonistic to each other; as Chaucer says, in his "Troilus and Cressida:"—

> "What might or more the sely lark say
> When that the sparhawke hath him in his foote?"

The male bird has a very graceful form and handsome plumage. The upper surface of the back and head of a dark-blue slate, with one small spot of white on the nape of the neck; the eyes orange; the chin, throat, and under-parts a reddish-brown, with dark transverse bars; legs and toes yellow. The female is a dark brown on back, &c., with the under-parts a greyish-white and barred.

THE KESTREL HAWK.

Another of the hawks, one of the long-winged species (the sparrow-hawk being short-winged), of the same order and family, is the KESTREL or WINDHOVER (*Falco tinnunculus*), often seen hovering over the meadow or marsh, its head being always pointed to the wind. It is a very harmless bird, preying chiefly on field-mice, frogs, beetles, and the like. Of course, it will occasionally take a small bird, as all hawks will when pressed by hunger, but this exception only proves the rule. It is well known that small birds will take no notice of the kestrel, but will con-

THE KESTREL HAWK.

tinue their songs, proving that they have no fear or consider themselves in any danger. Seebohm says: "Its presence is readily detected as it hovers in the air—

> 'As if let down from the heaven then
> By a viewless silken thread;'

now advancing towards you, flying upward some thirty feet above the earth, its wings flapping hurriedly or held perfectly motionless; now it is directly above you; you see its broad head turning restlessly from side to side; the wings seem in a perpetual quiver, and the broad tail is expanded to its fullest extent."

It is a very beautiful bird, and very graceful in its movements. In the male, the top of the head and neck a beautiful ash-grey, with long streaks; the back and wing-coverts fawn-colour, with small black spots; tail ash-blue grey; eyes dark brown; legs and toes yellow. The female is larger and more uniform in colour.

Willughby says: "The term kestrel is derived from the Greek word κεγρος, a millet, as if one should say millet bird," alluding to the mottled millet-like marks on the breast. The bird also goes by the name of the Stannel-Hawk, Stand-Hawk, or Steinfall. So Shakespeare, in "Twelfth Night," Act ii. sc. v. :—

"And with what wing the stanniel checks at it."

In Scotland we occasionally come across some of the grander species of the *Falconidæ*, as the GOLDEN EAGLE and the PEREGRINE FALCON. We were once fishing on Loch Coolin, in Ross-shire, when a splendid specimen of the former soared over our heads not very far above us, making for a neighbouring crag; and again in Argyleshire, whilst fishing the river Carnac, we put down our rod to watch two golden eagles on the opposite hill hunting for game (rabbits). One settled on a rock quite near, and with our binocular we got a splendid view of him and his eagle eye flashing in the sunlight.

CHAPTER IX.

THE BARN-OWL.

How often, when at eventide, on the banks of some stream, when the red sun has sunk beneath the horizon and the elms in the distance look almost black in the gloaming, we have stopped to watch the flight of the BARN-OWL (*Aluco flammeus*) as he searches with silent wing the water-meadows! How suddenly he will stop, hover for a moment, and swoop down into the long grass, and as suddenly rise again! He has missed his prey. Again he quarters his ground like a pointer; once more he is down—with success this time, for he rises with something dark in his claws—probably a young water-vole—and away he flies to yonder old ivy-mantled tower or ruined mill. We have scarcely made a few casts when he is again at work; he is on the other side now, gliding down the hedgerow. But if we watch him much longer we shall lose the chance at that big trout just flopped up near the opposite bank. We can see the rings he has made. Away goes the big alder into the midst of the round O's. Ah! we have him—whirr goes the reel. Look out for that low tree across the stream with its bed of tangled weeds. He means business; but, thanks to strong tackle and a judicious strain, we have turned him. The net is under him, and a fine two-and-half pounder lies on the dewy grass. But the owl? Oh! there he is. He must have gone and come again during the tussle. He has now got a much larger bunch in his claws—perhaps the old water-vole, or a rat from the hedgerow. What a friend of the farmers is this bonny owl; and yet how often, from ignorance and superstition, do we see this poor bird nailed to the barn-door or hanging to the gamekeeper's gallows-tree! Both farmer and gamekeeper have very strong opinions and murderous intentions about this bird. The one believes that it takes his young pigeons, the other that it destroys his pheasant chicks. Now, as the barn-owl seeks its food for the most part when the young pigeons and young pheasants are under the wings of their respective mothers, one does not quite see how this destruction can go on; it does, no doubt, occasionally pick up an unprotected bird or two, but its principal food consists of rats, mice, voles, frogs, and the like.

Seebohm says: "My friend, Frank Buckland, once found twenty dead rats in a barn-owl's nest, all fresh killed, and yet the stupid farmer will slay him if he can, under the delusion that he eats his pigeons. Out of between thirty or forty nests examined by Mr. Norgate, only in one instance did he find the remains of a bird. Out of 700 pellets examined by Dr. Altum, remains were found of 19 bats, 2513 mice, 1 mole, and 22 birds, 19 of which were sparrows. The barn-owl is one of the farmer's best friends."

The poor owls get scant justice from most country-folks. The barn-owl is the screech-owl, the dread of village boys and old women, who have a most superstitious awe of this bird. Should its screech be heard when watching the sick-bed, it bodes certain death. Spenser, in his "Faerie Queen," says:—

> "The messenger of death, the ghastly owl,
> With drery shriekes."

And Drayton:—

> "The shrieking stritch owl that doth never cry,
> But boding death."

Shakespeare, in "Richard III. :"—

> "Out on you, owls! nothing but songs of death."

In all countries where this bird is found it has a bad name with the ignorant and superstitious. In France, if a screech-owl shrieks on the chimney of a house where a woman is lying-in, a girl will be born, with ill-luck.

THE BARN OWL.

So Shakespeare, " Henry VI.," Part III., Act. v. sc. 6. :—

"The owl shrieked at thy birth, an evil sign."

In mythological history the screech-owl was once a man, Ascalphus by name, transformed by Ceres for his misdeeds into an owl, the messenger of approaching grief, a direful omen to mortals.

If there is light to see, stick your rod in the grass and put your binocular upon the bird as he slowly flies through the meadows. See, although he is called the white owl, and looks very white in the dusk, he is a yellow tawny-colour on the back of the head, altering to a tawny-greyish on the back; his wings beautifully barred with white spots; the breast and under-part a light pinkish-white; the facial and disc feathers round the eyes a silky white; black eyes; yellowish-pale beak, and black feet, covered with thick, short hairs.

We have mentioned the owl's *silent flight*. This arises from a most beautiful arrangement of the barbules of the wing-feathers (first pointed out by the late Mr. John Quekett), which renders their under-surface so soft as completely to deaden the sound which from the absence of such structure in the wings of other birds, is produced by the percussion with the air in the act of flight.

The barn-owl has many provincial names: Church-Owl, Hissing-Owl, Roarer, Billy-Wix, Woolert, Hoolet. In the Highlands it is *Gaillach-oidhche-gheal*, or "the white old woman of the night."

Some have asserted that the barn-owl hoots occasionally, but this is not the case. Its note is a weird kind of screech, very difficult to describe. Dresser says: "The call of the barn-owl is a loud, harsh, and most weird-sounding shriek; besides this, it sometimes makes a sound which is scarcely distinguishable from the snore of a man."

HEAD OF THE BARN-OWL.

THE BROWN OR TAWNY OWL.

The call-note of the BROWN or TAWNY OWL (*Syrnium aluco*) often greets our ears when returning homeward at dusk, one bird answering another from the neighbouring copses; their loud *Hōŏ-hōŏ-hōŏ* sounding so clear in the still air. It is chiefly in the autumn evenings that the voice of this owl is heard, generally just before dark and also in the early morning. In the breeding-season the male utters a most uncanny note, which makes one creep when heard in the silent forest-glades.

The dark, tawny, brown plumage sufficiently conceals the bird in the gloaming when seeking its food, which is

HEAD OF THE TAWNY OWL.

very varied—rats and mice, more especially field-mice, shrews, young hares and rabbits, fish, moles, large beetles, &c., &c. These owls are fond of frogs, and that frogs are well aware of this the following extract from the life of Edwards, the Scottish naturalist, clearly proves. After describing the noise made by a number of frogs on a moonlight night, he says: "Presently, when the whole of the vocalists had reached their highest notes, they became hushed in an instant. I was amazed at this, and began to wonder at the sudden termination of the concert. But, looking about, I observed a brown owl drop down, with the silence of death, on the top of a low dyke close by the orchestra."

This owl is not so dreaded as the barn-owl. In ancient times, amongst the Athenians, the brown owl was the bird of wisdom, and was associated with the goddess Minerva; Wisdom and Folly were often represented by an owl and a fool's cap and bells. Shakespeare says, in "Love's Labour's Lost:"—

> "When blood is nipped, and ways be foul,
> Then nightly sings the staring owl—
> Tu-whit, tu-who—a merry note."

And Walter Scott makes Goldthred sing a joyous song anent the owl:—

> "Of all the birds on bush or tree,
>   Commend me to the owl,
> Since he may best ensample be
>   To those the cup that trowl.
> For, when the sun hath left the west,
> He chooses the tree that he loves best,
> And he whoops out his song and laughs at his jest.
>   Then, though hours be late and weather foul,
>   We'll drink to the health of the bonny, bonny owl."

This owl has a number of provincial names—Tawny Hooting-Owl, Jenny-Howlet, Hoot-Owl, 'Ollering-Owl, Wood-Owl, Ivy-Owl, &c.

THE NIGHTJAR.

Should we happen, on leaving the river in the shades of evening, to cross on our way homeward a bit of ferny heath or dry field near a copse, our attention is aroused by hearing a peculiar snap, snap in the air, and a hawk-like bird will pass on its silent way close by us. This is the NIGHTJAR or DOR-HAWK (*Caprimulgus Europœus*). It has many other provincial names, as Fern-Owl, Night-Hawk, Churn-Owl, Evejar, Goat-Sucker, Puckeridge. All these names are derived from the supposed habits of the bird. Gilbert White says the country-people have a notion that the fern-owl, which they also call Puckeridge, is very injurious to weaning calves, by inflicting, as it strikes at them, the fatal distemper known to cow-leeches by the name of *Puckeridge*. It is probable that the cow-leeches of that day were very ignorant, and attributed this power to the bird to account for a disease which they knew nothing about.

In Italy the bird is accused of sucking goats; hence its name, *Caprimulgus*, or Goat-Sucker. It is needless to remark that all these supposed propensities have no existence. The bird lives entirely on insects, chiefly beetles. It is particularly fond of the dor-beetle—the shardborne beetle of Shakespeare—which often rushes by us of an evening with a loud hum; hence its name, Dor-Hawk. The peculiar churning, or rather croaking, noise is de-

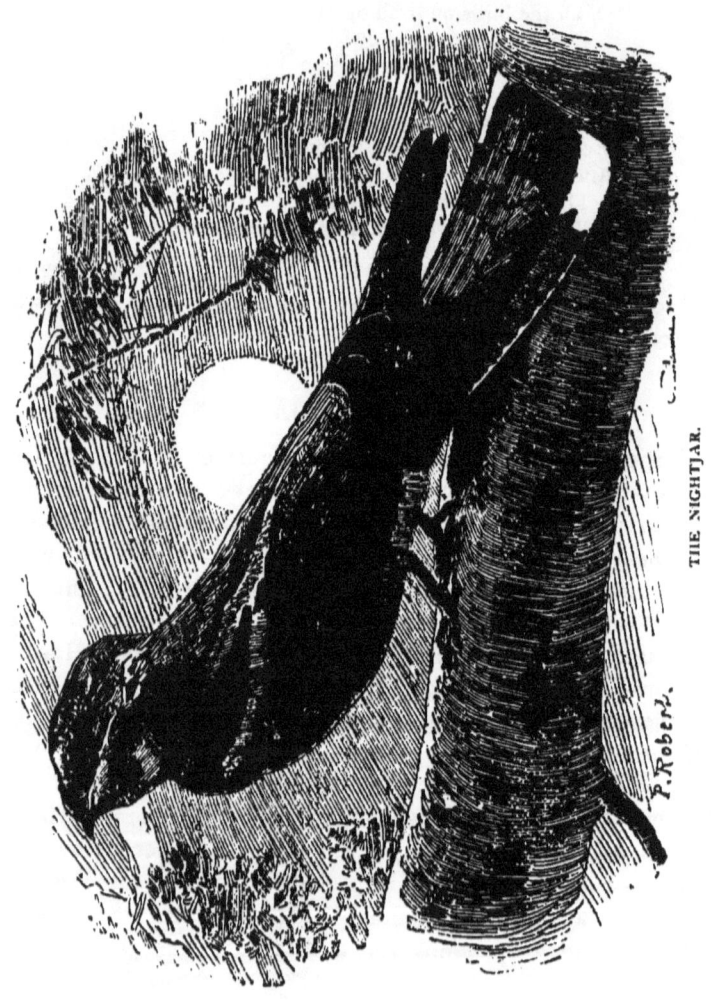

THE NIGHTJAR.

scribed by Wordsworth in his poem, "The Waggoner." This sound may be heard at a considerable distance:—

"The dor-hawk; solitary bird,
Round the dim crags on heaving pinions wheeling,
With untired voice sings an unvaried tune.
Those burring notes are all that can be heard,
In silence deeper far than that of deepest noon."

THE NIGHTJAR.

The burring note, however, is only uttered when the bird is at rest. Dresser ("Birds of Europe") says: "The whirring or churring note, something like that produced by a spinning-wheel, is uttered whilst the bird is perched on a branch; and when flying, it every now and again strikes the points of its wings together, making a sound which may be heard at a considerable distance, at the same time uttering a whistling note. It is by some observers said to possess the power of ventriloquism."

Macgillivray also states ("British Birds," vol. iii. p. 641) "that the whirring sound is made when at rest, and the whistling when on the wing." Harting, in "Birds of Middlesex," says: "I have heard the nightjar make a different noise on the wing, which sounds like '*Wh-ip, wh-ip;*' but I have not satisfied myself whether this sound proceeds from the bird's throat, or whether it is caused by striking the wings above the back as its flies."

A correspondent of the *St. James's Gazette*, writing about the nightjar's note, says there are three distinct and totally different notes: the burr, the cry, and the trill. As regards the latter, he says: "The trilling note is very peculiar. What I have seen is this: The two nightjars are perched each on a separate gable of the house; suddenly, first one, then the other, flies off with a loud slapping noise, produced by striking the wings together over the back. They sink slowly down with wings outspread, and alight upon the gravel-path under my window; they bow and sidle to each other, and then is heard a long-drawn and very musical bubbling note or trill, dying softly away into silence. I had often heard this sound in the distance and been puzzled by it, never dreaming that an amorous nightjar could produce any note so soft and liquid. I imagine it to be peculiar to the breeding season, and to be produced by the male bird only."

The bird, when seen close, is beautifully marked, and very like the grey markings of an oak-bough. When flushed it flies, if possible, to the nearest oak-tree, and there sits on a branch horizontally; and so like is the plumage to the bark, that it is most difficult to discover the bird unless you catch its dark lustrous eye. The middle toe or claw of each foot is curiously serrated—supposed to be used in combing the bristles on each side of the mouth when they get clogged with the hard wiry coverts of the beetles it preys upon.

There is a curious superstition in Nidderdale, Yorkshire, that these birds embody the souls of unbaptized infants doomed to wander for ever in the air, and are called Gabble-Ratchets—*i.e.*, Corpse-Hounds—a name equivalent to the Gabriel-Hounds of other localities, the unseen pack which is heard by night baying in the air—hence the Shropshire term for the bird, Uchfowl, or Corpse-Fowl. (Swainson's "Provincial Names of British Birds," p. 93).

HEAD OF THE NIGHTJAR.

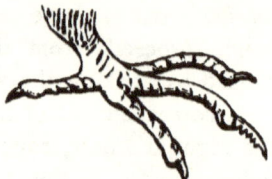

THE FOOT, SHOWING THE SERRATED CLAW.

The following interesting notes on this bird appeared in the *Zoologist*, August 1884 :—" Having had exceptional opportunities this summer, in Berkshire, of watching the habits of this curious bird, I venture to send the following remarks :—On its arrival, about the end of April or beginning of May, it is much bolder than it is later in the season. The note is loud and discordant then, and it is easy in the twilight to walk to the tree on which it may be sitting lengthwise on the branch, with head low. If disturbed it gives a peevish hoot, and claps its wings together behind, after the manner of some pigeons, pausing an instant after each clap to recover its equilibrium. Later on, as the breeding season approaches, its note becomes very ventriloquial, and it is then sometimes very difficult to stalk. The jarring note becomes much softer, and sometimes resembles the purring of a cat. If roused from its perch whilst making this noise, it continues the same note, letting it grow fainter till it dies away, and then gives the cry or hoot which is always uttered on the wing. The bird rarely appears in daylight, though I have heard one occasionally during the brightest summer afternoon. At half-past eight at midsummer they begin to appear, and continue till shortly after ten; but on moonlight evenings they may be heard till midnight. They lay earlier in the year than is generally supposed. On June 25 I found two 'nests' (?), each containing eggs; one lot had been sat on for some little time, and a keeper assured me that on June 19 he found young birds. When the hen is disturbed on her nest, if only one egg is laid, she flies straight away; but if she is sitting she will draw the intruder away by feigning a broken wing, at the same time uttering a cry of distress. The eggs are laid on the ground, not the slightest pretence of a nest being made; in one case, however, some fallen pine-needles had been scraped away till the bare soil was reached. The hen-bird appears to select her breeding-place some time before laying, notwithstanding the absence of a nest. When the eggs are reached there is difficulty in seeing them, so much do they resemble the ground; few better examples of 'protective mimicry' could be given, the eggs exactly matching in colour the leaden sand with its white stones found in the district of Berkshire where these notes were taken.—T. N. POSTLETHWAITE, Millom, Cumberland."

The nightjar is a summer migrant, arriving in May and departing in September.

CHAPTER X.

THE REPTILES.

DOUBTLESS St. Patrick, when he excommunicated all reptiles from Ireland (the only exception being in favour of the Common Lizard), supposed he was conferring a great favour on the Green Island. Let us be thankful that he did not come over here and do the same.

The Reptiles constitute a class of cold-blooded vertebrate animals. The term "cold-blooded" signifies "that the power of producing animal heat is so limited as scarcely to be appreciable, and not sufficient, therefore, to keep up any standard temperature of the body, nor to prevent it from following all the thermal variations of the atmosphere or water by which they are surrounded."

The two orders of the class Reptilia which come under our notice are the *Squamata (Sauria)*, family, *Lacertidæ*, the Lizards; with the *Saurophidia*, the Blind Worm, family, *Anguidæ*, and the (*Squamata Ophidia*), the Serpents, divided into the families *Colubridæ* and *Viperidæ*.

Of the class Amphibia, which Bell separates from Reptilia, two of the orders, *Anoura* and *Uradella*, are indigenous to this country. The *Anoura* include the *Ranidæ*, Frogs, and *Bufonidæ*, Toads; the *Uradella*, the *Salamandridæ*, the Newts.

THE LIZARDS.

Of the *Lacertidæ* or Lizards we have two distinct kinds, one in the genus *Lacerta*, and the other in that of *Zootoca*.

The first, the SAND-LIZARD (*Lacerta agilis*), Fig. A, is found chiefly about the sandy heaths in the neighbourhood

THE LIZARDS. 167

of Poole, Dorsetshire, but not at all common. It is *oviparous*, and deposits its eggs in hollows in the sand which it excavates on purpose, and having carefully covered them

over with a layer of sand, leaves them to be hatched by solar heat.

The colour varies: some are darkish-brown, others sandy-brown on the upper-parts, with a lateral series of black rounded spots each marked with a yellowish-white dot or line in the centre; some are more or less of a green colour on the sides. The total length is about 7 inches.

The second, the VIVIPAROUS LIZARD—the Common Lizard or Nimble Lizard—(*Zootoca vivipara*), Fig. B, is much more common. It is one of the few reptiles found in Ireland.

It is a very graceful little animal, and on sunny days can be seen on many a bank basking and enjoying itself. It is very rapid in its movements, and darts its tongue on its prey with unerring aim. It is particularly fond of dipterous insects, flies, &c. The female retains its eggs within its oviducts until the young are ready to leave them, and they are produced alive. It is, therefore, *ovoviviparous*.

This little lizard is smaller than *Lacerta agilis*. The colours and marking vary much. The general ground-colours of the upper-parts are greenish-brown, with a dark-brown line down the middle of the back, often somewhat interrupted; a broad fascia extends parallel with this on each side, commencing behind the eyes and extending to a greater or less length down the tail; between these and the former are often one or more rows of black dots, and similar ones occur in many individuals in the broad lateral fascia. The under side of the body and base of the tail in the male are bright orange spotted with black. In the female these parts, as well as the tail, pale greyish-green without spots (Bell). The usual length is from $5\frac{1}{2}$ to $6\frac{1}{2}$ inches.

THE SLOW-WORM.

Another reptile often wantonly killed, being ignorantly taken for a snake, is the BLIND or SLOW WORM (*Anguis fragilis;* family, *Anguidæ*); it is, in fact, closely allied to the lizards. Bell says: " It makes its appearance at an earlier season than any other of our scaled reptiles, and retires in the autumn under masses of decayed wood or leaves, or into soft dry soil, where it is covered with heath or brushwood, and penetrates to a considerable depth in such situations by means of its smooth, rounded muzzle and even polished body."

It is very harmless and inoffensive, and very timid; it has the power of contracting its muscles to such an extent as to become perfectly stiff, and then if at all roughly handled or struck with a twig or whip, will break into a number of pieces. Hence the name *Fragilis* given to this animal by Linnæus.

It is *ovo-viviparous*, and generally produces its young in June or July. It sheds its skin like the true snakes, leaving its slough turned inside out on the bushes.

HEAD OF THE SLOW-WORM.

It is about 10 to 12 inches in length, sometimes even much longer, about equal thickness, with a very small head, with very small eyes, brilliant and quick, with perfect eyelids, as in the lizards. Teeth very small, slightly hooked; tongue broad, notched at the extremity, but not bifid, as in the snakes; tail obtuse, about half the length of the body; general colour brownish silvery-grey, with several parallel longitudinal rows of black spots.

It is very fond of eating a little white slug, *Limax agrestis*, so very common in our fields and gardens, as well as insects and earthworms.

THE SNAKES.

Of the snakes, we have in this country only one venomous and one non-venomous. The COMMON VIPER or ADDER, Fig. 1 (*Pelius Berus;* family, *Viperidæ*), is far more common in Scotland than the common snake. Although the bite of an adder will cause very unpleasant symptoms, we believe there is no record of death being caused by it among the human race. We have seen dogs, especially pointers and setters, bitten by the viper in Scotland, and it appears to us that the dogs suffered more in the hot weather, when probably the virulence of the poison is more effective. The viper is *ovo-viviparous*, thus differing from the common snake, and produces as many as from twelve to twenty young ones at each birth. It hibernates in the winter months. The general colour varies considerably — in some olive or a rich brown, in others a dirty yellow-brown, in others again extremely dark, almost black. This variation of colour has led some to suppose that there were different species, but this is not the case. Adders vary much in length, but any over 24 inches must be considered as great rarities. The best remedy for the bite is ammonia, employed both externally and taken internally. The common form of sal volatile will do. A mark between the eyes, a spot on each side of the hinder part of the head, and a zigzag line running the whole length of the body and tail, as well as a row of small triangular spots on each side, all of a much darker colour than the ground of the body, easily distinguishes the viper from the RINGED or COMMON SNAKE, Fig. 2 (*Natrix torquata;* family, *Colubridæ*), which is plentifully distributed over all parts of England. It swims with ease and rapidity, and we have often seen this snake cross a river; its fondness for frogs as food may account for its resorting to moist places and streams. This snake is *oviparous*, and generally deposits its eggs either in a hotbed of leaves or in dunghills. It hibernates in winter. The upper parts of the body and head are of a light brownish-grey, with a green or olive tinge. Behind the head is a broad band of yellow, and close behind two broad spots of black. Two rows of small black spots are arranged alternately down the back, and larger ones at

FIG. 1.

FIG. 2.

the sides, which vary in size. The under-parts are pale-bluish, often marked with black.

The female is much larger than the male, and has been known to attain the length of 4 feet.

Snakes will also take small fish. We were sketching by the side of a lake, when suddenly a commotion in the water near us attracted our attention. We saw a snake had seized a small bleak, and was swimming towards the shore with it in its mouth. The rest of the shoal were following and surrounding the snake, as if inclined to attack it, but it got safely to some hole in the bank, and disappeared from view.

The question has often been asked whether adders swim. All snakes swim, and swim well, and most snakes are extremely fond of water.

THE FROGS.

In the class Amphibia, are included our frogs, toads, and newts, examples of which we may constantly come across in our rambles. We have two, if not three, species of frogs, the COMMON FROG (*Rana temporaria*), the SCOTTISH FROG (*Rana Scotica*), and perhaps the EDIBLE FROG (*Rana esculenta*). We all know the young of the frogs—the tadpoles—which infest the stagnant waters, &c., in the summer. The tadpole of the frog may be distinguished from that of the toad by it, being of a considerably lighter colour, the toad tadpole being almost black. The frog is distributed over all parts of our island, wherever there is a pool or river, or even wherever there is sufficient moisture necessary to preserve the respirable condition of the skin—respiration being carried on both by the lungs and the skin. The food of the frog consists of various insects and small slugs, and on account of this both frogs and toads are very useful in gardens. Bell says: "The manner in which the frog takes its food is very interesting. As in the toad, the tongue is doubled back upon itself when at rest; and being imbued with a viscous secretion at the extremity, it is suddenly thrown forwards upon the insect, which, being caught by the adhesive matter upon it, is instantly drawn into the mouth by the sudden return of the tongue to its former position, and is then swallowed."

The frog hibernates in the winter, generally in the mud at the bottom of the water. Here they may be found congregated in masses so closely together as to form a continuous heap, thus preserving an equable temperature and securing themselves from external injury. In the spring they separate and emerge, and at once proceed to the reproduction of their species. At this time a peculiar

THE FROG.

temporary development of a warty protuberance or knob takes place on the thumbs for the purpose of close adhesion ; and so powerful is this instinct of adhesion, that instances have been known of frogs attaching themselves to fish, generally to the eyes, without the possibility of the fish shaking the reptile off. In the "Complete Angler" Walton gives an example of this, in Chapter VIII., fourth day ; and although Dubravius, a bishop of Bohemia; who wrote a book of "Fish and Fishponds," saw what happened with his own eyes, it is scarcely credible. Pennant, in his "Zoology," vol. iv. p. 10, says : "As frogs adhere closely

THE FROGS.

to the backs of their own species, so we know they will do the same with fish. Walton mentions a strange story of their destroying pike; but that they will injure, if not eventually kill, carp appears evident, from the following relation : 'A very few years ago, on fishing a pond belonging to Mr. Pitt, of Encomb, Dorsetshire, great numbers of carp were found each with a frog mounted on it; the hind-legs clinging to the back and the fore-legs fixed in the corner of each eye of the fish, which were thin and greatly wasted, teased by carrying so disagreeable a load.' These frogs we imagine to be males disappointed of a mate."

Taverner ("Experiments on Fish, &c.," printed in 1600) says: "In the moneth of March, at what time todes doe ingender, the tode will many times covet to fasten himselfe uppon the heade of the carpe, and will thereby invenime the carpe in such sort that the carpe will swell as great as he may hold, so that his scales will stand as it were on edge, and his eyes stand out of his head neare halfe an inch, in very ugly sort, and in the end will for the most part die thereof, and it is dangerous for any person to eate of any such carpe so invenimed."

Bell ("British Reptiles," p. 91), after quoting Walton's anecdote, says: "I have often heard my father relate an instance of a similar fact, though with somewhat more adherence to the simple truth of the case. As he was walking in the spring on the banks of a large piece of water at Wimpole, the seat of Lord Hardwicke, he observed a large pike swimming in a very sluggish manner near the surface of the water, having two dark-coloured patches on the side, which he thought must be occasioned by disease. A few days afterwards he saw the same pike floating dead upon the surface of the water, and having drawn it to land by means of a stick, he found that the dark-coloured masses which he had observed on the former occasion were two living frogs, still attached to the fish, and that so firmly that it required some force to push them off. There can be no doubt that the diseased state of the pike facilitated the approach and adhesion of the frogs, to which they were primarily impelled by the sexual instinct."

The change in colour in the frog arises from variations in temperature, effects of light or absence of the same, and other causes, one of which is fear, which is exhibited in a great degree when the frog is under the influence or in the jaws of the common snake.

The *Rana Scotica* is comparatively rare, and has been confounded by authors with the *Rana esculenta*, or edible frog, but it is, we believe, a question whether this latter is indigenous to this country.

THE TOADS.

Of the Toads there are two British species. One, the COMMON TOAD or PADDOCK (*Bufa vulgaris*), is often met with in our rambles. It is an inoffensive, harmless creature, much despised and often killed from ignorant prejudice on the supposition that it is venomous. It can be easily

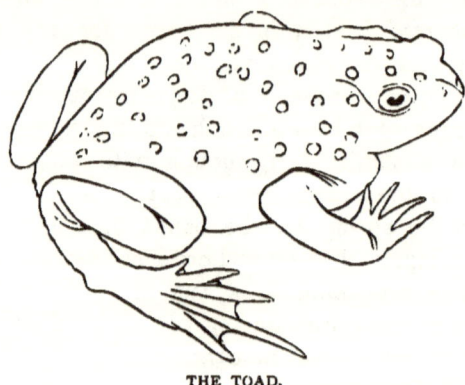

THE TOAD.

tamed, and is susceptible of considerable attachment to those who treat it with kindness. Pennant in his "British Zoology" (Appendix) gives a very interesting anecdote in relation to this.

In olden days the toad, or a part of the animal, was a necessary ingredient in the witches' potions, be they administered to the love-sick or for baser purposes.

THE TOADS.

Ben Jonson in his witches' charms gives the following:—

> "The scritch owle's eggs and the feathers black,
> The bloud of the frogge and the bone in his backe,
> I have been getting, and made of his skin
> A purset to keep Sir Cranion in.
> I went to the toad, who breeds under the wall,
> I charmed him out, and he came at my call,
> I scratch'd out the eyes of the owle before,
> I tore the bat's wings, what would you have more?"

That eminent naturalist, the late Thomas Bell, in his work on the "British Reptiles," article "Toad," says: "Few animals have ever suffered more undeserved persecution as the victims of an absurd and ignorant prejudice than the toad. Condemned by common consent as a disgusting, odious, and venomous reptile, the proverbial emblem of all that is malicious and hateful in the human character, it is placed under universal ban, and treated as an outlaw both by man and boy throughout the country. Should I be able, by the following history of its habits and manners, to show that it is, on the contrary, highly useful, perfectly harmless, inoffensive, and even timid, and susceptible of no inconsiderable degree of discriminating attachment to those who treat it with kindness, it is hoped that some few individuals may be thus rescued from those barbarous acts of cruelty to which the species is almost everywhere subjected."

There is no doubt that the toad does produce a highly nauseous secretion from its skin which is most unpleasant to other animals, but it is not venomous in the usual acceptation of the term.

The second species is the NATTER JACK (*Bufo calamita*), not so common as the previous, but still found in many places; it was formerly found in considerable numbers about Blackheath and Deptford. It is not so sluggish as the common toad. The eyes are much more elevated above the surface of the head, and more prominent. It is more terrestrial in its habits, and the rudimentary sixth toe found in the common toad is absent. It is also easily distinguished by the yellowish line along the middle of the back.

THE NEWTS.

Of the Newts, we have four British species, all of which are common in our ponds and stagnant waters.

The COMMON WARTY NEWT or GREAT WATER-NEWT (*Triton cristatus*), Fig. 1, is well distinguished by its dorsal crest, which becomes very prominent on the male in the spring. The female lays its eggs, one at a time, in the following manner:—She selects some leaf of a water-plant, sits upon its edge, and folding it by means of her hinder-feet, deposits a single egg in the duplicature of

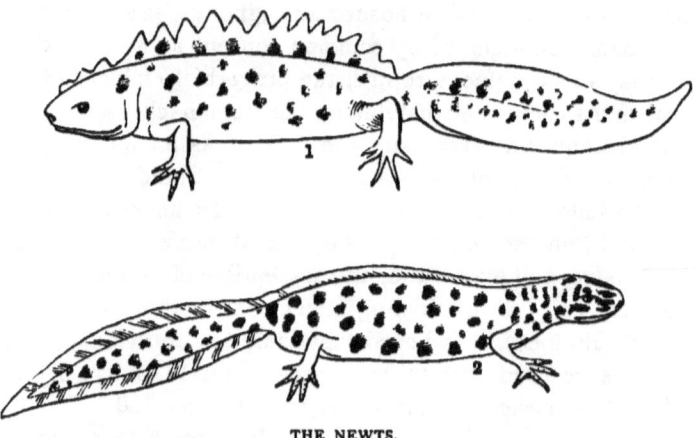

THE NEWTS.

the folded part of the leaf, which is glued securely together, and the egg effectually protected. She then quits the leaf, and after a short time goes through the same process on another leaf, and so on, laying altogether a considerable number of eggs. The food of this newt consists of aquatic insects and small living animals, such as worms and tadpoles, and the like. The colour is a black-brown or yellow-brown on the upper-parts, with dark round spots; the under-parts bright orange-red, with round black spots; sides of the tail in the males of a shiny pearl-white. In the breeding season the male puts on the deep indented crest, which extends the whole length of the back; it is separated from the crest of the tail by an interval, without any elevation; it disappears during the winter.

The STRAIGHT-LIPPED WATER-NEWT (*Triton Bibronii*), is not often found.

The COMMON SMOOTH NEWT, EFT, or EVET (*Lissotriton punctatus*), Fig. 2, is the most common; distinguishable by its small size and smoothness, all warts and tubercles being entirely absent. It is found in almost every ditch and pond, and is often eaten by the first-described species and by different kinds of fish. Its own food comprises aquatic insects (especially in their larva state), small aquatic worms, and water-snails. They will take flies that settle on the surface of the water, and will also eat the tadpoles.

The word eft or evet is probably derived from the Anglo-Saxon *ef-an*, "smooth," from the smoothness of its skin. The colour varies very much. The male is brownish-grey in the upper-parts, yellow beneath, which in the spring becomes a bright orange, marked with numbers of round dark spots. The crest in the breeding season is often tipped with bright red or violet. The female is of a light yellow-brown, the under-parts quite plain.

The PALMATED SMOOTH NEWT (*Lissotriton palmipes*) is much larger than the previous species, and differs in the form of the upper lip, which is pendulous at the sides, and the spots which cover the body are more numerous and smaller. The colour varies considerably.

CHAPTER XI.

THE FISHES.

"As in successive course the seasons roll,
So circling pleasures recreate the soul.
When genial spring a living warmth bestows,
And o'er the year her verdant mantle throws,
No swelling inundation hides the grounds,
But crystal currents glide within their bounds;
The finny brood their wonted haunts forsake,
Float in the sun, and skim along the lake,
With frequent leap they range the shallow streams,
Their silver coats reflect the dazzling beams."
—GAY, *Rural Sports.*

BLAINE ("Encyclopædia of Rural Sports;" article, "Fishing") says: "*Man we believe to have been first frugivorous,* and then carnivorous, and next piscivorous, which left him what he now is—omnivorous. The methods he first employed to abstract the finny tribes from the waters around him were, without doubt, rude, and consequently only partially successful; but, as these improved by practice, fishing at length became a most important, and finally a most interesting, pursuit. As an amusement, it calls forth considerable powers of invention and much dexterity of operation. As an economic art it has become most important, by giving rise to vast national fisheries, which have proved objects of great interest in the policy of every civilised country bounded by the sea. To improve and extend these has been the aim of the wisest statesmen; to defend them has nerved the arm of the bravest warriors; and to acquire a general and particular knowledge of the creatures which they are meant to capture has been the study of the most enlightened philosophers."

THE FISHES.

Before describing the various fishes, we would say a few words on those organs by which their mental intelligence and powers of movement are regulated.

Naturalists and physiologists up to a very recent period have placed the fishes into the lowest class of vertebrate animals as regards their mental capacities. Professor Rymer Jones, writing some thirty years ago, says: "We are justified in anticipating that in intelligence, and in the relative perfection of their senses, fishes should be less highly endowed than the other vertebrate classes;" and he continues in another paragraph: "With such inferiority in their powers of communication with the external world, and with faculties so circumscribed, we might justly infer that, as relates to their intellectual powers, fishes hold a position equally debased and degraded."

The observations of later years, owing to a much closer investigation of their habits—particularly in fresh-water fish—not only show that they have considerable mental intelligence, but that they are able to exercise the powers both of reason and deduction.

The brain of a fish differs very considerably in capacity from the higher vertebrata, and occupies but a small portion of the cranial cavity, the interval between the cranium and the brain being much less in young fish than in adults, which proves that the brain does not grow in the same proportion to the rest of the body; indeed, it has been observed by Cuvier and others, the size of the brain is nearly equal in individuals of the same species, even although the body of one is twice the size of that of the other. The brain of a fish consists of several masses placed one behind the other, either in pairs or singly; these masses may be regarded as so many distinct ganglia, the complexity and perfection of which are gradually increased as we proceed upwards in the scale.

The cerebral hemispheres of all vertebrated animals are supposed to be the seat of the mental powers. In the higher animals, in quadrupeds, and more especially in man, the proportionate size of the hemispheres of the brain is so large as to conceal the ganglia and other parts;

but in the lower forms these relative dimensions become smaller and the structure less complicated, until in fishes, which are supposed by some to be the least intelligent of this great division of the animal kingdom, they are found in such a rudimentary state that they are frequently inferior in size even to the olfactory or optic ganglia of the higher forms. Yet, with all this, some fish show a great amount of intellectual discrimination—some more than others. A trout, for instance, which always feeds near the surface, soon learns the difference between the shadow or the substance of man and that of other vertebrates. It will, when feeding, take no notice of any animal by the side of the stream or the shadow of a passing cloud; but let a man, or his shadow—or even the shadow of a rod—pass near it, the fish will at once be scared—either rush madly away, or, in fishing parlance, will be put down. It is the same, only in a less degree, with all; but as some remain for the most part in deep water, they are less alarmed than those which have to find their food near the surface, and which appear to be able to discriminate between those which can harm them and those which cannot; and, like all *feræ naturæ*, become more wary the more they come in contact with man, the destroyer of them all.

As the vision of fish is of great importance to their safety and welfare, the construction of the eye is worthy of consideration.

The second pair of ganglia in the brain of a fish give origin to the optic nerve, each nerve being composed of a broad band, folded up like a fan, and enclosed in a dense membrane. The eye itself differs in many points from that of terrestrial animals. "Its organisation," says Professor Rymer Jones, "being, of course, adapted to bring the rays of light to a focus upon the retina in the denser element in which the fish resides, the power of the crystalline lens is therefore increased to the utmost extent, and the anterior-posterior diameter of the eyeball necessarily contracted in the same ratio, in order that the retina may be placed exactly in the extremely short focus of the powerful lens.

THE FISHES.

The eyes of all vertebrate animals are constructed upon principles essentially similar, and have, as a rule, the same parts as are found in the human eye, and only modified according to circumstances. The following short description of the eye in man will assist in understanding that of the fish. The globe—or, as it is more generally termed, the ball—of the eye is composed of membranes placed one within the other, and of humours or fluids which they enclose. The membranes are the *conjunctiva* which lines the free border and inner surface of the eyelids. The *sclerotic*, a fibrous membrane, firm and resistant in texture, extending from the entrance of the optic nerve to the border of *cornea*, which occupies the anterior fifth of the globe of the eye. The anterior surface is convex and prominent, is in contact with the conjunctiva; the posterior is concave, and is lined by the membrane of the aqueous humour. The *choroid membrane*, vascular and composed of minute arteries and veins, united by cellular tissue, lies between the sclerotic and the *retina*, which is placed between the choroid membrane and the vitreous humour—soft and pulpy in its structure, and transparent in the living subject. The iris, or coloured circle, seen through the transparent cornea, resembles a partition to divide the interval between the cornea and lens into two parts; the interval is filled with the *aqueous humour*, and the iris moves freely in it. The space between it and the cornea is called the *anterior chamber*, which is the larger; that behind, the *posterior chamber*. The *aqueous* humour is a thin pellucid fluid, filling up the two chambers of the eye, occupying the space between the cornea and *crystalline lens*, which is placed at the union of the anterior third with the two posterior thirds of the eye behind the iris, and embedded in the *vitreous humour*, which fills up the posterior two-thirds of the globe of the eye. It consists of a thin transparent fluid, enclosed in a membrane, called the hyaloid membrane. This structure, although thin and transparent, is firm, particularly in the forepart.

That which strikes the attention when examining the

eye of a fish is the size of the crystalline lens and its spherical form. "This shape," says Rymer Jones, "and the extreme density of texture which the lens exhibits, are indeed perfectly indispensable. The aqueous humour, being nearly of the same density as the external element, would have no power in deflecting the rays of light towards a focus, and consequently the aqueous fluid in fishes is barely sufficient in quantity to allow the free suspension of the iris; the vitreous humour, for the same reason, would be scarcely more efficient than the aqueous in changing the course of rays entering the eye, and hence the necessity for that extraordinary magnifying power conferred on the lens." But the focus of the crystalline will be short in proportion as its power is increased; every arrangement, therefore, has been made to approximate the retina to the posterior surface of the lens: the eyeball is flattened by diminishing the relative quantity of the vitreous humour, and a section of the eye shows that its shape is very far from that of a perfect sphere. This flattened form could not, however, have been maintained in fishes, had not a special provision been made for the purpose in the construction of the sclerotic. The outer tunic of the eye, therefore, generally contains two cartilaginous plates embedded in its tissue, which are sufficiently firm in their texture to prevent any alteration in the shape of the eyeball. The vitreous humour and crystalline lens in many fishes are kept *in situ* by a ligament placed for the purpose. In some fishes, as the salmon, this ligament is of a dark colour.

Another peculiarity in the eye of the bony fishes is the presence of a vascular organ at the back of the eyeball, interposed between the choroid tunic and a brilliant metallic-coloured membrane, which invests the choroid externally. This organ is of a crescent form, and always of a deep-red colour, principally made up of bloodvessels, always much ramified, forming a vascular network in the choroid. The nature of this organ is uncertain; some suppose it to be muscular, others glandular. Professor Rymer Jones thinks it an erectile tissue analogous to that of the *corpus caver-*

nosum, and that it has some influence in accommodating the form of the eye to distances or to the density of the surrounding medium.

The presence of this choroid gland, a "*rete mirabile*," says Günther, is always combined with that of a pseudo branchia. Bony fishes, in which this latter is absent, have no choroid gland.

The pupil of the eye in fishes is always very large, so as to take in as much light as possible; but it is generally without motion and unable to contract. Mr. Forster, however ("Scientific Angler"), states that he has seen the pupil contract. Six muscles serve to turn the eye in various directions—four recti and two oblique—as in man; and although, as Günther says, in the range of their vision and acuteness of sight, fishes are very inferior to the higher classes of vertebrates, yet, at the same time, it is evident that they perceive their prey or approaching danger from a considerable distance, and some are able to discriminate with clearness and precision one colour from another—at least, we may suppose so when we see how often the salmon family reject, in the artificial flies placed above them, one colour and instantly take another.

It would be an interesting study to determine at what angle fish perceive objects behind them or directly in front of them. It would appear in many that the vision is chiefly directed upwards and laterally, but how far their vision, particularly in fish which get their food chiefly on the surface of the water, extends backwards and forwards is not yet determined. An object placed laterally or above will almost immediately attract or scare. We have many times been able to get close to a trout by approaching directly from behind, when the slightest deviation laterally would send him away. In approaching fish, not sufficient consideration is given as to the powers of refraction and the medium through which a fish sees; and it must be remembered that fish do not see objects as we see them.

Mr. George M. Kelson, who has, perhaps, more than any one, studied the habits of the salmon, says: "I am fully convinced that salmon see the fly when it is in a position exactly over their heads. Actual observation soon taught me that much. Often have I been looking through binocular glasses and seen salmon behind perpendicular rocks take notice of flies properly presented immediately above them, the fisherman having payed out his line for the purpose. What happens is this: if the fly is one they mean taking, the dash after it is so instantaneous and sudden that I have never once been able to see exactly what they do. But if the fly is one they merely inspect—and they invariably do so unless they have been pricked or overthrashed—they turn their heads sideways towards it, so as to get a better view with one eye. So far as can be judged from practical experience, I am also convinced that salmon are apt to discriminate with astonishing precision one coloured fly from another; and, in my opinion, they readily detect the least variation in the composition of a fly. It must be well known to old and observant hands that the small difference in size of a startling feather, such as that of the jungle-fowl, affects the fish, and as frequently influences their movements."

There is another organ in fish which, having a great deal to do with the movements of many of them, is worth a few moments' consideration—the swimming or air bladder. This is placed beneath the spine, and is firmly bound down by the peritoneum. The outer coat is generally very strong, and composed of a peculiar fibrous substance, which, when obtained from some fishes, is the isinglass of commerce. It is lined internally with a very delicate and thin membrane. The shape varies in different families. In the perch it is a simple sac closed at both extremities. In the carps it is divided into two portions—an anterior and posterior—with a narrow constricted communication between the two. But whatever may be the shape, it serves a specific purpose, viz., to alter the specific gravity of the fish, so that it may rise or sink in the water. By simply compressing this bladder by approximating the walls of the abdomen, or by means of a muscular apparatus provided for the purpose upon a principle with which every one is

THE FISHES.

familiar, the fish sinks in proportion to the degree of pressure to which the contained air is subjected, and as the compressed air is again permitted to expand, the creature, becoming more buoyant, rises towards the surface.

In many fish, the perch, &c., the air-bladder is closed and there is no escape for the confined air; and in those fish with this form of bladder which live at great depths, the very bringing them up to the surface, the air or gas no longer being compressed by the weight of the water, bursts the swimming-bladder. This is often seen in fishing for cod, &c.

Müller says: "The air of the swimming-bladder is not derived from without. It is secreted by the inner surface of the sac. The proportion which the oxygen bears to the nitrogen is sometimes greater and sometimes less in the air of the swimming-bladder than it is in the atmosphere. In the air of the swimming-bladder of fishes in inland lakes Ermann found but a small proportion of oxygen. From the air of the air-bladder of fishes which live in the depths of the ocean, Biot obtained from 69 to 87 per cent. of oxygen, while the air with which the water of the ocean at a considerable depth was impregnated contained only 29 parts of oxygen and 71 parts of nitrogen in 100 parts. The composition of the air varies, however, even in the same species of fish. In spring and summer the air is said to contain less oxygen than in autumn; and sometimes it contains no oxygen. According to the mean result obtained from a great number of experiments, the air of the swimming-bladder of the carp would consist of 71 parts oxygen, 52 parts of carbonic acid, and 377 parts of nitrogen in 1000 parts. In some fishes, as in the carp, the air-bladder communicates with the pharynx. In many fishes this communication does not exist at all, and in them there is usually a red vascular and peculiar tissue in the walls of the bladder destined for the secretion of the air, and even when the air-bladder does communicate with the pharynx, it is most probable that the air is secreted."

It is hardly necessary to say that fish breathe by means of their gills; but fish absorb oxgyen and exhale carbonic acid not only by their gills, but by the whole surface of their body as long as they are surrounded with water impregnated with atmospheric air; and the respiration of a great many fishes will continue in the air and will absorb oxygen if the gills are kept constantly moist. So that, as Müller observes, aquatic respiration differs from respiration in the air less essentially than might at first appear. A moist surface is necessary even in pulmonary respiration.

A few words as regards the fins—organs of the utmost importance in fish. Günther divides the fins into *vertical* or *unpaired* and *horizontal* or *paired*. He says: "Any of them may be present or absent, and their position, number, and form are most important guides in determining the affinities of fishes."

The vertical fins are the dorsal, caudal, and anal.

Very important differences are to be found in the dorsal fin, which is either spinous (*acanthopterygian*) as in the perches, &c., or soft-rayed (*malacopterygian*) as in *salmones*, &c. It may be single or divided.

The caudal fin is seldom symmetrical, *i.e.*, the upper half is not equal with the lower.

The anal fin may be single or plural, long or short, or entirely absent; and in some (the *acanthopterygians*) the foremost rays are spined.

The horizontal or paired fins consist of two pairs, pectorals and ventrals.

The pectoral fins are the homologues of the anterior limbs of the higher Vertebrata, and are always inserted immediately behind the gill-covers.

The ventral fins are the homologues of the posterior extremities, and inserted in the abdominal surface, either behind the pectorals as in the Salmon, or below them as in the Red Mullet, or in advance of them as in the Burbot.

"The fins," says Günther, "are organs of motion, but it is chiefly the tail and the caudal fin by which the fish impels itself forward. To execute energetic locomotion the tail and caudal fin are strongly bent with rapidity alternately towards the right and left; whilst a gentle motion forward is effected by a simply undulating action of

THE FISHES.

the caudal fin, the lobes of which are like the blades of a screw. Retrograde motions can be made by fish in an imperfect manner only by forward strokes of the pectoral fins. When the fish wants to turn towards the left he gives a stroke of the tail to the right, the right pectoral acting simultaneously, whilst the left remains adpressed to the body. Thus the pectoral fins assist in the progressive movements of the fish, but rather directing its course than acting as powerful propellers. The chief function of the paired fins is to maintain the balance of the fish in the water, which is always the most unsteady where there is no weight to sink it. When the pectoral of one side or the pectoral and ventral of the same side are removed, the fish loses its balance and falls on the side opposite; when both pectorals are removed the fish's head sinks; on removal of the dorsal and anal fins the motions of the fish assume a zigzag course. A fish deprived of all fins, as well as a dead fish, floats with the belly upwards, the back being the heavier part of the body" ("Study of Fishes," p. 44).

As we find that these so-called low order of vertebrates have the power of exercising some of the higher functions of those much further advanced in the scale of creation, and as it is well known that all animals of the higher orders are obliged to have a period of repose for the brain, the question arises, Have fish the power of sleeping?

Müller ("Elements of Physiology") says: "The excitement of the organic processes in the brain which attends an active state of mind gradually renders that organ incapable of maintaining the mental action, and thus induces sleep, which is to the brain what bodily fatigue is to other parts of the nervous system. The cessation or remission of mental activity during sleep, in its turn, however, affords an opportunity for the restoration of integrity to the organic conditions of the cerebrum, by which they regain their excitability. The brain, whose action is essential to the manifestation of the mind, obeys in fact the general law which prevails over all organic phenomena—that the phenomena of life, being particular states induced in the organic structures, are attended with changes in the constituent matter of those structures. The duration of the periodical state of sleep, and the time at which it occurs, are dependent partially on external and partially on internal causes.

"The causes determining duration of sleep and waking are seated, however, in the organism itself. The periodical recurrence of sleep and the waking state is, therefore, essentially connected with something in the nature of animals, and is not dependent on the simple alternations of day and night. The sleep of animals is a phenomenon dependent on a change in the animal part of the organism alone. All the functions of organic life—namely, the processes ministering to nutrition, with all the involuntary movements attending them—pursue their ordinary course. Even the involuntary movements of the animal system of muscles, such as those of respiration, and many other movements of the same kind, do not partake of the repose of sleep. The organic system has its period of remission and rest, but these are not coincident with the sleep of animal life, and are very different for different organs.

"All the phenomena of organic life, and indeed all the phenomena presented by the animal body, with the exception of the true animal functions which are under the influence of the mind, obey a law of absolute necessity, and the nutrition and maintenance even of the organs of animal life are not dependent on the operations of the mind or intellect.

"We may, therefore, regard sleep and the waking state as the result of a species of antagonism between the organic and the animal life, in which the animal functions, governed by the mind, from time to time become free to act, while at other times they are repressed by the organic force acting in obedience to a law of creative nature.

"Sleep, in a greater or less degree, as Aristotle correctly remarked, falls to the share of all animals. Some even dream—the dog, for instance, barks in his sleep. In some the periods of sleep are less distinct and regular; and this is particularly the case in the cold-blooded animals. They, however, appear to be subject to states analogous to sleep.

THE FISHES.

Frogs, which croak a part of the night during summer, become quiet after midnight, especially when the pairing season is past.

"Insects and spiders are often found in a lethargic or torpid state; and it is probable that all animals in which no regular periods of sleep and waking have hitherto been observed have an equivalent for sleep in the state of inactivity and rest which they from time to time present."

Dr. John Davy ("The Angler and his Friend," p. 74) says: "As to sleep, fish, I believe, do rest and steep their senses in forgetfulness, and this probably more by night than by day. I am led to this conclusion from observations on very young fish. These, about the darkest part of the night, when confined in glasses so as to be easily seen, I have noticed immovable, resting on the bottom of the vessel, and quite regardless of an approaching object, not stirring until they have been almost touched; so different in this respect from their manner when awake by day. Perhaps their hours of rest vary in some measure according to their age and wants; probably old and large fish, like the larger carnivorous beasts—the lion, the tiger—rest by day and then sleep; we know that night is their principal feeding-time."

The late Mr. Manley ("Notes on Fish and Fishing, p. 15) asks: "Do fish sleep? It may be presumed, to start with, that they do, otherwise they would form an exception to all other vertebrate animals. I need hardly say that the fact of their having no eyelids to close would be no bar to profound sleep." At p. 16—"Query: Do they swim in their sleep as somnambulists walk in their sleep? But as to the fact whether they sleep or not at night, or whether they take an occasional nap or siesta by day, what is the evidence? No one, as the proverb suggests, ever 'caught a weasel asleep;' and I do not know of any one who ever caught a fish asleep. A friend of mine, a good angler and ardent naturalist, adduces as evidence of fish sleeping the fact which, he says, he has established by experiment and observation—viz., that for above six hours during the night in winter, and about two in summer, no sound of fish moving is to be heard, and none are to be taken by any bait, with the exception of eels, which are clearly nocturnal in their habits. He says, also, that he has constantly taken chub with a white moth all through the early part of the summer's night, but that the fish have suddenly ceased rising just two hours before dawn, and have remained quiescent until the dawn had quite broken."

We think that fish would be more often found asleep if anglers' observations were directly applied for the purpose of ascertaining this. We ourselves, on two occasions, have caught fish asleep. Once, on the Kennet, we, with a friend, noticed a large trout lying near the bank, a little out of the stream of a shallow; as we came up to him, although a bright day, he never stirred. We got close to him, threw a fly over him, of which he took no notice. We threw the shadow of ourselves and rod over him; no movement. We then got exactly opposite to him, so close we could count all his spots, and he appeared as if he was steadily looking at us. His fins were acting gently all this time. We again threw our shadows over him, waved our rods and arms; he never moved. We then, with the aid of the rod, gently touched him, and it was not till after rather a strong push did he wake up and dart with great velocity across the stream.

Precisely the same thing happened to us on the Test, with a much larger fish lying near the bank. We put our shadows over him, came up alongside, admired his shape, size, and colour; and it was not until we poked him with our rod did he take any notice, and then he rushed across the stream, making a great wave.

The question has often been asked, " Do fish feel pain on being hooked?" The answer has generally been given in accordance with the wishes of the respondent. A correspondent of the *Fishing Gazette*, says: " Please calm the conscience of the 'Amateur Angler.' Fish wriggle when on the hook because they want to get away. They do not and cannot suffer any pain—because they've got no sensory nerves. They have frequently told me so."

THE FISHES.

The late Mr. Manley says: "It is a question which must often suggest itself to the angler, and many must have wished that they could unhesitatingly answer it in the negative. I think they may do so." One reason he gives is the fact that numerous instances of the same fish being caught immediately after being previously hooked; but that only proves that the power of memory is deficient, which, considering the size of a fish's brain, can be easily understood.

One has only to study the nerve distribution in the head of a fish to see at once that relatively fish must feel pain when hooked—in what degree is another question. The nerves which supply the whole of the head in the fish are the same as those which are distributed in the head of most of the vertebrates. Professor Rymer Jones says that, "with the exception of the ninth pair (the hypoglossal nerve), which are not met with in fishes, both in distribution and in number the nerves of fishes precisely accord with those of man. In fishes the vagus and fifth pair are usually the largest, and the fifth pair are the sensitive nerves supplying the orbit. The parts about the nose, the upper and lower jaws, indeed as far as regards the face the distribution is exactly similar to that in man; but in fishes it also gives off, branches to the gill-covers to the top of the skull, joining a large branch of the eighth pair, and issuing from the cranium through a hole in the parietal bone, passes along the whole length of the back on each side of the dorsal fin, round twigs from all the intercostal nerves, and supplying the muscles of the fins and fin-rays." From this description of the distribution of this sensitive nerve, it is impossible to say that fish do not feel pain, but it is probably very evanescent.

The fact is, it is difficult to follow any sport comprising the capture of animals without producing pain more or less, and it is well to bear in mind Sir Humphry Davy's remarks, that there is danger in analysing too closely the moral character of any of our field-sports. "If," he says, "all men were Pythagoreans and professed the Brahmin's creed, it would be undoubtedly cruel to destroy any form of animal life; but if fish are to be eaten, I see no more harm in capturing them by skill and ingenuity with an artificial fly, than in pulling them out of the water by main force with the net; and in general, when taken by the common fishermen, fish are permitted to die slowly and to suffer in the air for want of their natural element, whereas every good angler, as soon as his fish is landed, either destroys its life immediately, if it is wanted for food, or returns it into the water."

CHAPTER XII.

THE fishes about to be described are, with one exception (the Lampreys), to be found among the sub-class TELEOSTEI or BONY FISHES. That is, the skeleton is ossified; the vertebræ (spinal column) completely formed; the branchiæ (gills) free, and protected by a bony gill-cover.

This great sub-class is again divided into six Orders; each order subdivided into families, genera, and species.

THE PERCH.

The first Order is that of the ACANTHOPTERYGII (from two Greek words, *acantha*, a thorn or prickle, and *pteros*, a fin), distinguished by having a portion of the dorsal, anal, and ventral fins not articulated, forming spines; the air-bladder, when present, being completely closed and having no air duct. To this order—

The family *Percidæ* (the Perches) belong. Günther says: "In the *Percidæ* the scales extend rarely over the vertical fins; the lateral line is generally present, and continuous from the head to the caudal fin. All the teeth are simple and conical. There are no barbels, and no bony stay to the preoperculum."

The FRESH-WATER PERCH (*Perca fluviatilis*) is found in a great many of the rivers, lakes, and ponds in England. It is more rare in Scotland. Yarrell, however, says that it is found in some of the lochs of North Britain, and it is pretty generally distributed in nearly all the fresh-waters of Ireland.

Day says the name of the fish is of Greek origin, signifying "dark colour," which probably refers to the dark bands across the body; "while this nomenclature has been introduced with but little change into almost every European country."

Aristotle describes the perch under the name of περκζ; the Roman name was *Perca;* in Italy it is called *Pergosa;* in Prussia, *Perseke;* in France, *La Perche;* in Germany, *Barsch*. In England this fish has various provincial names, as Barse in Westmoreland, Base in Cumberland, Trasting in Cheshire, Crutchet in Warwickshire, &c.

Perch are gregarious, are fond of deep holes, and are

THE PERCH.

found often by the sides of piers or bridges, or the wooden or brick camp sheathings in mill-dams, &c. They often form companies, and herd together, the smaller separately from the larger fish:—

> "Pearch, like the Tartar clans, in troops remove,
> And urged by famine, or by pleasure, rove;
> But if one pris'ner, as in war, you seize,
> You'll prosper, master of the camp with ease;
> For, like the wicked, unalarm'd they view
> Their fellows perish, and their path."
> —*The Angler.*

Francis, however, says: "Where they are at all fished, there are few fish more capricious or careful in biting than large perch; small ones may often be taken in numbers, but not so when they gain experience."

A perch in season is a very beautiful fish. It has a bright olive-green back, with its lighter sides tinged with yellowish-golden pink, and getting brighter below. The transverse dark bands, so characteristic, set off the rest; there are generally five, but occasionally six are found, the first commencing below the first spine of the first dorsal fin, and the sixth—if it is present—is just in front of the tail. The first dorsal fin is greyish-brown with two black spots; the second dorsal is light brown, the pectoral fin brown; other fins of a bright vermilion. The irides are of a golden-yellow.

The perch has three cæcal appendages. The intestinal tract is the length of the fish without the tail. The air-bladder is large and simple, and thin in texture. The number of fin-rays are as follows:—First dorsal, 15, all spinous; second dorsal, 14, 1 spinous; pectoral, 14; ventral, 6, 1 spinous; anal, 10, 2 spinous; caudal, 17.

Deformed varieties of this fish are at times met with. Linnæus tells us of a hunch-backed perch found in Sweden, which goes by the name of Rudaborre. Pennant mentions one of the same kind in Cheshire, and Yarrell states that similar perch are found in Llyn Raithlyn, in Merionethshire, also mentioned by Pennant and Daniel, some of these deformed fish being above 2 lbs. in weight. Daniel also says that at Malham Water, not far from Settle, in Yorkshire, the perch grow to 5 lbs. and upwards, but that these large fish are all blind with one or both eyes.

Perch are very sluggish in cold weather, and are supposed to be injuriously affected by thunder and by frost. When spring sets in, or even in February if mild, they become more active, and will bite freely at almost any bait. It is stated on good authority that a perch has been taken with a hook baited with its own eye. Moses Browne says that "the curious remark, he seldom bites till the mulberry-tree buds, and until extreme frosts are passed."

The perch is said to be a slow grower, but this, we think, must depend on locality and food. Mr. Manley ("Fish and Fishing") says: "In answer to the question, 'To what size will an English perch attain?' I answer 4 lbs." But there are records of much greater weight. Pennant mentions one taken in the Serpentine of 9 lbs., Colonel Montagu records one of 8 lbs., and Mr. Jesse one of 5 lbs. 10 oz. One of 8 lbs. was caught in Dagenham Breach by Mr. Carter, as mentioned in Daniel's "British Rural Sports." A perch, however, nowadays, of from 2 or 3 lbs. is considered a very heavy fish. Those of the river Kennet are particularly fine and large, taken both at Hungerford and Newbury, and other localities. The late Mr. Francis Francis records a day's perch-fishing at Hungerford, where with a friend he basketed thirty-seven perch, weighing 60 lbs.

Perch spawn in April and May, depositing the ova in long riband-like bands on the leaves of water-plants, &c. A writer in the *Fishing Gazette* says: "The development of the embryo is very interesting, and is very rapid also. Very soon it may be seen to move, and when this stage is reached it is almost in constant motion, jerking itself spasmodically, and this goes on until it finally bursts its shell and is free. When about half incubated the eyes are distinctly visible, and it is a curious sight to watch the constant movement of a large number of embryos; move or shake them, and they at once become quiescent for a time." It was supposed at one time that the perch was hermaphrodite. Mr. Manley, *l. c.*, says: "I believe our own perch, if not every species of perch, is thus bisexual. It is," he says, "a remarkable fact, or what seems to be a fact, that nine out of every ten perch an angler takes are females, judging at least from the roe."

Yarrell, vol. i., first edition, p. 10, in describing the smooth serranus, one of the marine family of perches, says: "One peculiarity of the serrani must not be passed over. Cavolini and Cuvier have, after repeated examinations, described the smooth serranus and some other species of the genus as true hermaphrodites, one portion of each lobe of roe consisting of true ova, the other part having all the appearance of perfect milt, and both advancing to maturity simultaneously. A structure of a different kind, which must be considered accidental, has been observed by others in the carp, perch, mackerel, cod, whiting, and sole." But the perch (*Perca fluviatilis*), although the proportion of males to females appears to vary in different localities, must not be regarded as an hermaphrodite fish.

Buckland examined two females, one 3 lbs. 2 oz., the other 2 lbs. 11 oz. In the first he found 155,620 ova; in the other, 127,240; and Mr. Manley states that in a single half-pound fish no less than 280,000 eggs were found.

Blane says: "A curious fact is recorded, which would lead to a belief that these fish are migratory. It is asserted that all the Irish lakes and the Shannon river became stocked with them at the same time; but," he adds, "we know not on what authority the account rests."

THE POPE OR RUFFE.

The POPE or RUFFE (*Acerina vulgaris;* genus, *Acerina*) is closely allied to the perch. Yarrell says that the ruffe

THE POPE OR RUFFE.

is common to almost all the canals and rivers of England, particularly the Thames, the Isis, and the Cam. It was first discovered in England by Dr. Caius in the river Yare, near Norwich, who termed it Aspredo, owing to the roughness of its denticulated scales; hence also its name of Ruffe. Daniel ("Rural Sports") says the river Yare, in Norfolk, affords, perhaps, the greatest plenty. The pope, like the perch, is gregarious, assembling in large shoals and keeping in the deepest part of the water. It is a free biter, but never grows to any size; seldom above seven inches in length. Daniel says: "The ruffe, for its delicacy and richness of its flavour, as well as for its being considered very nourishing, is more admired than the perch;" and Blane says ruffe forms a good table treat broiled in buttered paper. Although it somewhat resembles the perch, it can be at once distinguished from that fish by the dorsal fin being continuous, the eye being brown, with a blue pupil, whereas in the perch the dorsal fin is divided into two, as it were, and the eye is yellow. The prevailing colour of the upper part of the body and head is olive-brown, yellowish-brown on the sides, silvery-white lower down; the lateral line prominent and strongly marked; a pearly-green tinge over the gill-cover; small brown spots are spread over the back, dorsal fin, and tail, being somewhat like bars in the latter. The other fins are pale brown.

Fin-rays: pectoral, 13–14; ventral, 6, 1 spinous; dorsal, 26, 14 spinous; anal, 7, 2 spinous; caudal, 17-18.

Day states that "a cruel custom obtains near Windsor, on the Thames, of pressing a cork tightly down on to the spines of the dorsal fin, and subsequently the fish is returned to the river." This is termed "plugging a pope," the origin of which is unknown. In some localities the fish is called a Jack Ruffe, in others Tommy Bass; the French call it *La Gremille Commune*.

THE BASSE.

The BASSE (*Labrax lupus;* genus, *Labrax*). Although we can hardly call the basse a fresh-water fish, yet at certain seasons it ascends a considerable way up some of our rivers and estuaries, and affords capital sport to the rod-fisher. This fish was well known to the ancients.

THE BASSE.

The Greeks call it *Labrax;* the Romans, *Lupus;* and these conjoined give it its specific name, *Labrax lupus.* It is found along the whole line of our southern coast, and is common in the Bristol Channel, where the fishermen often call it the Sea Salmon. It is a strong, active, and very voracious fish, living in shoals, comes out of the deep sea in May, and returns to it in October, and is therefore seldom met with in winter months. It is fond of frequenting wooden piers, and off the pier at Herne Bay many a good fish has been taken with a sand-eel or lug-worm, or a spoon-bait. The fishermen there call it the Sea-Dace, and sometimes the White Salmon. Gorlestone Pier and Britannia Pier at Yarmouth are favourite angling stations for this fish.

The body of the fish is elongated as compared with

THE BASSE.

that of the perch, and in shape resembles the salmon. The teeth are uniform in size, short and sharp, villiform in the jaw; the outer row, in the upper, somewhat larger than the rest, in an almost crescentic form on the vomer; in a band on the palatines and also at the base of the tongue.

The colour on the back of a darkish-grey, silvery on the sides and below. Dorsal, anal, and caudal fins greyish; pectoral and ventrals yellow-white. Fin-rays: dorsal, 9, 2nd 13, 1 spinous, which occasionally is absent; pectoral, 16; ventral, 6, 1 spinous; anal, 24, 3 spinous; caudal, 27.

In Scotland this fish is known as the Gape-Mouth; at Belfast, as the White Mullet or King of the Mullets; in France it is *Le Bass* or *Le Loup;* in Holland, *Zee Bass.*

THE RIVER BULL-HEAD.

The River Bull-Head or Miller's Thumb (*Cottus gobio, Cottidæ*) receives, according to Yarrell, its first name from the bigness of its head as compared to the rest of the body, and the second from its likeness to a miller's thumb, which in former days was used in a peculiar manner for the purpose of testing the quality of the meal, whereby the

THE MILLER'S THUMB.

thumb became broad, flat, and smooth, and exactly like the shape of the fish's head. Yarrell states that he derived this information from the late John Constable, R.A., "whose father, being one of the large Suffolk millers, was early initiated in all the mysteries of that peculiar business."

This fish is a very undesirable inhabitant of a trout-stream, for although it occasionally chokes a dabchick by sticking in its gullet, yet it is a great devourer of ova and fry, and is never satisfied.[1] It should be exterminated wherever possible. It generally frequents shallow streams and hides itself under stones, darts quickly on its prey, but is a bad swmimer.

The bull-head has a variety of provincial names. It is

---

[1] "Dabchick and Bull-head.

"Sir,—I found a dead dabchick in the stream here the other day, choked by a large bull-head, and am having it set up in defunct attitude by Mr. Peachey, of Elizabeth Street, Pimlico, as a monument of feathered vice.—I am, &c.,

"C. R. C. H.

Slough."

THE STICKLEBACKS.

the Bull-Knob of Derbyshire, the Cull of Gloucestershire, the Codpole of Bucks and Berks, the Nozzle-Head of Hants, Tommy Logge of Kent. In France it is called *Chabots*.

The-fin rays are—1st, dorsal, 6–9; 2nd, 17 or 18; pectoral, 15; ventral, 3; anal, 13; caudal, 11, all spinous.

The mouth is wide, jaws nearly equal, in which are many sharp teeth, as well as on the fore-part of the vomer; the irides yellow; pupils dark blue; on the preoperculum there is one spine, which is curved upwards. The dorsal fins are united by a membrane. All the rays of the fins are spotted. Colour of the body brownish-black; sides lighter, with small black spots; under-surface of the head and belly white. Spawns March and April.

THE STICKLEBACKS.

The STICKLEBACK, *Gasterosteus*, *Gasterosteidæ*. Yarrell makes six species: the three-spined, rough-tailed stickleback, *Gasterosteus aculeatus;* the half-armed, *G. semi-armatus;* the smooth-tailed, *G. leiurus;* the short-spined, *G. brachycentrus;* the four-spined, *G. spinulosus;* and the ten-spined, *G. pungitius*.

Day ("The Fishes of Great Britain and Ireland") gives only two species: (1) *Gasterosteus aculeatus*, the THREE-

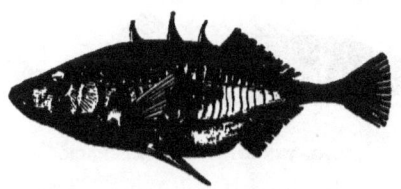

THE THREE-SPINED STICKLEBACK.

SPINED STICKLEBACK, and (2) *Gasterosteus pungitius*, the TEN-SPINED STICKLEBACK; and makes six varieties: (1) *G. trachurus;* (2) *G. semicoloratus;* (3) *G. semiarmatus*, distinguished by the want of arming by defensive plates along the sides of the line—a very common variety; (4) *G. gymnurus*, the lateral plates extending no farther than the end of the ray of the pectoral fin; (5) *G. brachycentrus*, with very short spines, found chiefly in Ireland; and the four-spined, *G. spinulosus*.

The late F. Buckland ("Familiar History of British Fishes") says these active and greedy little fish are extremely destructive to the fry of other species. Mr. Baker writes: "A banstickle which I kept for some time did on the 4th of May devour in five hours' time seventy-four young dace, which were about a quarter of an inch long, and of the thickness of a horse-hair. Two days after, it swallowed sixty-two, and would, I am persuaded, have eaten as many every day could I have procured them for it." All who have watched the stickleback in an aquarium know that this fish is a nest-builder, in which the female deposits its eggs, spawning some time from April to June, and nothing can exceed the attention of the male in watching over the nest and its contents. An interesting account of this habit of the stickleback is given by Buckland. It appears that in a nest built on April 23 the fry first appeared on May 21.

A large number of sticklebacks are sometimes caught in the lower portion of the Thames, in the whitebait-nets.

The TEN-SPINED STICKLEBACK (*G. pungitius*) is one of the smallest fishes, rarely exceeding two inches. It is found in great shoals in the spring. It generally has ten spines

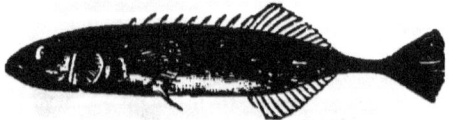

THE TEN-SPINED STICKLEBACK.

(hence its name) in front of the dorsal fin, but at times only nine, and sometimes even reduced to eight. The general colour is a yellowish or olive green on the back; sides and belly silvery-white, with minute specks of black; fins pale yellowish-white.

The COMMON STICKLEBACK (*G. aculeatus*) seldom exceeds two and a half inches in length. It is found in almost every brook, river, and lake, and occasionally descends in large shoals to the estuaries and to the sea. It is extremely active

THE STICKLEBACKS.

and pugnacious. Yarrell says it rarely lives above two years. It has a most voracious appetite, and feeds upon worms, larvæ, water-shrimps, and even upon its own eggs, as well as the ova and fry of every kind of fish. Its pugnacity is extraordinary, and it does not hesitate to attack fish many times bigger than itself. Day quotes the following:—" Mr. Mable, at the Weston-super-Mare Museum, had some three-spined sticklebacks, and some roach (*Leuciscus rutilus*) were also added. With this invasion the prior inhabitants were dissatisfied, but not frightened, as they forthwith attacked the new-comers, biting at them anywhere until they became thoroughly cowed. These little tyrants were observed to place themselves in front of the roach, steady themselves by their tails, and then suddenly dart straight at the lips of their intended prey, from which they bit pieces out. These attacks were continued until all the roach had been killed, when they were eaten by their conquerors. Mr. O'Delly mentions placing some carp and tench in a tank containing sticklebacks. Almost immediately these little furies attacked the carp, and gave them no rest until they died, which occurred in three or four days, not one of them having more than a vestige of fin or tail left. The tench were left alone" ("Fishes of Great Britain," p. 242).

The male is the fighting fish, and is easily distinguished from the females by its brilliant green, red, and pink colours, especially at spawning-time.

The stickleback has a variety of provincial names—Prickleback, Tittlebat, Jacksharp, Sharpling, Pricky, Bansticle, Fiery-Loch (male), Enemy-Chit (female), Stickling, Stanstickle, Pow, &c. In Scotland, Banstickle and Sharplin. In Ireland, Thornback and Pinkeen. In France it is called *L'Épinoche*.

The stickleback is a favourite food with trout. Only the other day we landed a fine trout of one and a half pound, and we took no less than twelve of these little fish from its maw. In some of the Scotch lakes where sticklebacks have been introduced the trout have thriven wonderfully upon them. Where fish were formerly considered large at a pound-weight, they are now taken from 2 to 4 lbs.

CHAPTER XII.

A LARGE proportion of our fresh-water fish belong to the fourth order of the bony fishes, the PHYSOSTOMI (from the Greek words *phuse*, a bladder, and *stoma*, a mouth); *i.e.*, the air-bladder, if present, has a pneumatic duct. All the fin-rays are articulated; sometimes the first of the dorsal and pectoral fins are ossified. Ventral fins, if present, are abdominal and without spines.

Dr. Günther ("Introduction to the Study of Fishes") divides this order into thirty-one families, of which the third (*Cyprinidæ*), the tenth (*Esociæ*), the fifteenth (*Salmonidæ*), and the thirty-first (*Murænidæ*) claim our particular attention.

The third family—*Cyprinidæ*—Günther says, has the "body generally covered with scales, head naked, margin of the upper jaw formed by the intermaxillaries, belly rounded—or, if trenchant, without ossifications; no adipose fin, stomach without blind sac, pyloric appendages none, mouth toothless, lower pharyngeal bones well developed, falciform, sub-parallel to the branchial arches, provided with teeth which are arranged in one, two, or three series; air-bladder large, divided into an anterior and posterior portion by a constriction, or into a right or left portion enclosed in an osseous capsule; ovarian sacs closed."

Day ("Fishes of Great Britain and Ireland") in his definition of this family says: "A single, rayed, dorsal fin; ventrals inverted posterior to the pectorals, head scaleless, body scaled or scaleless, never covered by osseous plates; no *cul-de-sac* to stomach, no pyloric appendages; air-bladder, if present, large; it may be divided by a constriction into an anterior and posterior portion, neither of which are enclosed in bone (*Cyprinidæ*), or into two lateral

THE CARP.

portions partially or entirely enclosed in a bony capsule (*Cobitidinæ*)."

This large family is found in most of the fresh-waters of the Old and New World. Day remarks "that in Europe and Asia it appears to take the place of the *Salmonidæ* of the colder North, increasing in numbers as the tropics are approached, until in Asia the genera are very numerous. Some few forms are said occasionally to be found in salt lakes; also some are stated to descend to the brackish tidal waters of estuaries, or even to the sea."

This family is also found in a fossil state in the Tertiary fresh-water formations, and many of these can be referred to existing genera.

In the *Cyprinidæ*, some are omnivorous, others carnivorous, insectivorous, or graminivorous; Day, *l. c.*, says "the difference in their food being indicated to a great extent by the character of their interior pharyngeal teeth and the extent of their digestive tract."

First group, *Cyprinus*.—The anal fin very short, with not more than five or six (exceptionally seven) branched rays; dorsal fin opposite ventrals. Abdomen not compressed; lateral line running along the median line of the tail. Mouth frequently with barbels, never more than four in number. Pharyngeal teeth generally in a triple series in the Old World genera; in a double or single series in the North American forms, which are small and feebly developed. Air-bladder present without osseous covering (Günther).

We shall divide our British *Cyprinoids* into two groups—(1.) Those with barbels, as the Common Carp (*Cyprinus carpio*), the Barbel (*Barbus vulgaris*), the Gudgeon (*Cyprinus gobio*), and the Loaches (*Cobitidinæ*), which form a second group of the true carps. (2.) Those without barbels—the Crucian Carp (*Carassius vulgaris*), the Bream (*Abramis brama*), the Roach (*Leuciscus rutilus*), the Chub (*L. cephalus*), the Rudd (*L. erythropthalmus*), the Dace (*L. vulgaris*), the Bleak (*L. Alburnus*), the Minnow (*L. phoxinus*), and the Tench (*Tinca vulgaris*).

THE CARP.

The CARP (*Cyprinus carpio*) is supposed to have come originally from the East, and then transported to Germany and Sweden. Its first introduction into England, according to Sir Richard Baker, "was about the year 1514:"—

> "Hops and turkies, carps and beer,
> Came into England all in one year."

Daniel, in "Rural Sports," gives a different version :—

> "Turkies, carps, hops, pickerell and beer,
> Came into England all in one year."

Hawkins (fourth edition "Complete Angler") gives yet another :—

> "Hops, Reformation, turkies, carps and beer,
> Came into England all in one year."

The carp, however, is mentioned in the "Boke of St. Albans," printed by Wynkyn de Worde in 1496, as "a dayntous fishe." Turkeys were not known in England till 1524, and hops somewhat later, as in 1528 the Parliament petitioned against their use as a "wicked weed." In 1504, according to Daniel, at the Inthronisation Feast of Archbishop Warham, we read of "carpe in sharpe sauce" and "carp in armine," whatever that means. Beer was drunk in England in 1422.

Yarrell states that "carp are said to have been introduced into Ireland in the reign of James I."

Gilbert White, in his fortieth letter to Thomas Pennant, September 2, 1774, says: "In the garden of the Black Bear Inn, in the town of Reading, is a stream or canal running under the stables and out into the fields on the other side of the road. In this water are many carps, which lie rolling about in sight, being fed by travellers who amuse themselves by tossing them bread; but as soon as the weather grows at all severe these fishes are no longer seen, because they retire under the stables, where they remain till the return of spring. Do they lie in a torpid state? If they do not, how are they supported?"

Yarrell says they probably eat little or nothing during winter, and are supposed to bury themselves in the mud. Their food in summer consists of insects, worms, and the softer parts of aquatic plants. They are said to be particularly partial to lettuce-leaves.

Hepell says, that in winter, carp huddle together in the mud in concentric circles, with their heads together.

The carp is a very wary fish. Moses Browne remarks that he is called the River-Fox. Izaak Walton says the carp is "a stately, a good, and a subtle fish." Sometimes it requires a considerable amount of patience to catch a carp; at others the fish will bite freely. The Rev. Thomas Weaver, in his commendatory verses in the second edition of "Walton's Complete Angler," has the following lines, which were omitted in all the later editions:—

> "And there the cunning carp you may
> Beguile with paste; if you'll but stay
> And watch in time, you'll have your wish,
> For paste and patience catch the fish."

Carp grow to a great size and live to a great age. The largest ever recorded is one mentioned by Brōck, which was taken near Frankfort-on-the-Oder, weighing 70 lbs., and being nine feet in length. Mrs. Garrick told Hawkins that she had seen in Germany the head of a carp served up at table big enough to fill a large dish. The largest carp recorded in this country was one killed at Pain's Hill, Cobham, which weighed 26 lbs.

Some of the carp at Fontainebleau are said to be 200 years old, and the custodians show some which they say are white from age; but Buckland states that this colour is due to disease, and not to age.

In the carp the mouth is small, a barbule at the upper part of each corner of the mouth, with a second smaller one above, on each side; teeth pharyngeal; nostrils large; eyes small; gill-covers marked with striæ, radiating from the anterior edge; nape and back rising suddenly.

The dorsal fin has 21 to 25 rays; pectoral, 15 to 16; ventral, 8; anal, 7 to 8; caudal, 20, forked; the caudal rays of the two halves of the tail always unequal.

The body is covered with large scales; the general colour olive-brown, head darkest; belly yellowish-white; lateral line interrupted and straight; irides golden-yellow; fins dark-brown.

Day ("The Fishes of Great Britain and Ireland," p. 158, vol. ii.) says: "It has been observed that the fishes which afford the best evidence of a ruminating action are Cyprinoids, as carp, tench, and bream, when peristaltic movements occur in the alimentary canal, and successive regurgitations of the contents of the stomach induce actions of the pharyngeal jaws as the half-bruised food comes in contact with them, and excites a succession of swellings and subsidences of the irritable palate as portions of the regurgitated food are pressed against it. On the occiput, behind the roof of the palate, is a single grinding tooth or plate, which has opposed to it the two inferior pharyngeal bones armed with teeth; while in the front of this plate, and forming the roof of the palate, is a thick, soft, vascular, and highly sensitive mass, which becomes thinner anteriorly, and is believed to be useful in taste, being supplied by branches of the glosso-pharyngeal nerves. This mass is commonly known as the 'carp's tongue,' and held in great esteem among epicures."

THE BARBEL.

The carp spawns in May and June, depositing its ova among aquatic plants. Its breeding powers are very great. As many as 700,000 eggs have been found in one female, and that not of the largest size; indeed, the name *Carp* is derived from the Greek *Kupris*, Cyprus, where Venus was first worshipped, and may have been given, as Dr. Day says, to this fish in order to symbolise its extraordinary fecundity.

THE BARBEL.

Günther says: " No other genus of *Cyprinoids* is composed of so many species as the genus of Barbels, about two hundred being known from the tropical and temperate parts of the Old World; it is not represented in the New World. We have only one species in Great Britain, viz., the Barbel (*Barbus vulgaris, Cyprinus barbus*)."

"The barbell," says Gesner, "is so called from, or by reason of, his beard, or wattles at his mouth, his mouth being under his nose or chaps; and he is one of the leather-

THE BARBEL.

mouthed fish that has his teeth in his throat. He loves to live in very swift streams."

The barbel is distinguished from the carp in having the dorsal and anal fins short; with a strong deviated bony ray at the dorsal fin; the mouth furnished with four barbules, two near the point of the nose and one at the angle of the mouth, on each side.

Buckland, in his " Curiosities of Natural History," thus writes of this fish :—"The barbel has not a tooth in his head; his mouth is made for poking about among the stones at the bottom of the rivers and procuring his food, which consists of almost anything. He is a regular freshwater pig, and lives by picking up what he can find, be it animal or vegetable."

In this country they sometimes, but very rarely, are found as heavy as 15 lbs.; a barbel of 10 lbs. is a rare fish. On the Danube and other large rivers on the Continent they reach a much greater size.

Day ("Fishes of Great Britain," &c.) says: "It is a matter of history that subsequent to a dreadful carnage between the Turks and the Austrians on the banks of the Danube, barbels were found of such vast size and in such numbers as to become a subject of record; while, their propensity to human flesh being well known, the circumstance was attributed to the heaps of dead bodies which had been thrown into the river."

Some barbel frequent the bottom of still rivers and their deeper parts; others are found in the rapids below the weirs and in swift streams.

The barbel is a handsome fish; of a greenish, olivaceous colour, with a gold bronzy burnish; belly white; eyes golden; dorsal and caudal fins brownish, tinged with red; pectoral, ventral, and anal fins flesh-red; lateral line nearly straight. It spawns May and June, the spawn being placed, not on aquatic plants, but on the gravel, which is subsequently covered over by the parent fish with great care.

The fin-rays are: dorsal, 12–13; pectoral, 18; ventral, 9–10; anal, 7–8; caudal, 20.

The flesh is very unpalatable and coarse; although Dame Juliana Berners says that it is "a sweet fysshe." Even old Izaak is obliged to say that "he is not accounted the best fish to eat, neither for his wholesomeness nor his taste; but the male is reputed much better than the female, whose spawn is very hurtful."

Day says: "The Jews, having boiled them with vinegar and oil, eat them not only at their white feasts, but whenever they can obtain them."

Barbel at times produce sickness and diarrhœa in some

THE GUDGEON.

who have eaten of its flesh. In the "Boke of St. Albans" it says: "The barbyll is a sweet fysshe, but it is quasy meete and perrylous for manny body."

Barbel-fishing is a favourite pastime of some Thames anglers, and fish of from 6 lbs. to 8 lbs., to even 10 lbs., are at times taken, and give very fine sport:—

> "How pleasant in a dog-day sun,
> When all on land looks dry and dun,
> To spend the day upon the river,
> On whose banks the osiers quiver;
> In a punt for barbel fishing
> (Or anything not worth dishing),
>   With 'merrie companie'!"

Salter states that in 1816 he knew of a barbel in Hampton Court deeps that had several times broken away from the hooks, and was supposed to weigh about 30 lbs. From his bold and piratical practices he was nick-named Paul Jones.

Barbel are omnivorous. The name is derived from *barba*, a beard. They are prolific breeders, and after spawning frequent the swiftest piece of water they can find.

THE GUDGEON.

The GUDGEON (*Gobio fluviatilis, Cyprinus gobio*).

Izaak Walton says: "The gudgeon is reputed a fish of excellent taste, and to be very wholesome. He is of fine

THE GUDGEON.

shape, of a silver-colour, and beautified with black spots on his body and tail." There is a small barbel at the angle of the mouth; pharyngeal teeth hooked at the end.

The gudgeon is found in many rivers, but the Thames, Mersey, Colne, Kennet, and Avon produce the finest. It is gregarious, and feeds on worms, aquatic insects, larvæ, small mollusca, and the ova and small fry of other fish. It is, therefore, not a desirable fish in trout-streams.

The colour of the upper part of head, back, and sides olive-brown, spotted with black; eyes orange-red, pupil large and dark; gill-covers greenish-white; all the under surface of the body white, tinged with brown; dorsal fin and tail pale brown, spotted with darker brown.

Fin-rays: dorsal, 9–10; pectoral, 15–16; ventral, 8–9; anal, 8; caudal, 19.

The fish is considered a great delicacy, and gudgeon fishing on the Thames is an institution, and to some, as Daniel says, a singular fascination. He relates a story of the minister of Thames Ditton, who was engaged to be married to the daughter of the Bishop of London on a certain day, but he went out fishing for gudgeon that morning, and, unfortunately for him, overstayed the canonical hour, and the lady, greatly offended at his neglect, refused to have anything more to say to him.

Old writers say that gudgeon spawn twice or thrice a year. The late Mr. Manley also, in "Fish and Fishing," remarks that they increase marvellously, spawning, it is said, no less than three times in the year; and it is a curious fact in natural history that the females outnumber the males by six to one. Dr. Day says: "They spawn in April, May, and June, depositing their small bluish ova among stones and in the shallow streams. In some localities they are supposed to breed three times a year, or take a long time spawning." Yarrell says the gudgeon spawns in May, and he found its fry an inch long in August.

THE LOACHES.

The LOACHES (*Cobitidinæ*). The head small; body long; there are six barbels, all on the upper lip; the pharyngeal teeth in a single series; and the air-bladder partly or entirely enclosed in a bony capsule.

Day says: "The Loaches, or the second group of the true Carps, are composed of forms in which the two lateral portions of the air-bladder are partially or entirely enclosed in bone, enabling it to be employed as an acoustic organ. These are mostly small fishes, and the two genera represented in Great Britain may thus be recognised:—

"With an erectile spine near the orbit, *Cobitis*.
"Without any, *Nemacheilus*."

The SPINED LOACH (*Cobitis tænia*), the Groundling, according to Yarrell, is rather a rare fish; but more probably its diminutive size causes it to escape observation. It is found in the Kennet and the Trent, as well as in some of the streams in Cambridgeshire.

The body is more elongated and more compressed, particularly behind the dorsal fin, than in the common loach. Six small barbels hang from the upper lip; the mouth and eyes are small; pectoral fin long and narrow. Colour of a lightish-brown; two rows of dark spots run along the back and sides, and another row, larger and more conspicuous, below the lateral line; a dark spot on the base of the caudal fin.

The fin rays are: dorsal, 8; pectoral, 9; ventral, 7; anal, 6; caudal, 15; all having dark bands.

The COMMON LOACH (*Nemacheilus barbatula, Cobitis barbatula*), also called the Loche, Groundling, Stone Loach, Beardie, Groundbait, Lie Still, Tommy Loach, &c., is found in almost all clear streams running over gravel, hiding itself under stones. It is very voracious, eating everything in the shape of larvæ, ova, insects, and the like.

The body is elongated, smooth, covered with a mucous secretion, rounded in form before the dorsal fin, and compressed posterior to it. The dorsal fin commences half-way between the point of the nose and the end of the fleshy part of the tail; the mouth small, and placed underneath; it has six barbels, four over the upper lip and two at each angle; the head, body, and sides are clouded and spotted with brown on a yellowish-white ground; abdomen whitish-yellow; all the fins spotted or banded with dark brown; the dorsal fin and caudal most so, the latter with a large dark blotch; the eyes (irides) blue.

The fin-rays: dorsal, 9; pectoral, 12; ventral, 7; anal, 6; caudal, 19; vertebræ, 36.

Yarrell mentions two peculiarities in the structure of the bones in the loach: "Attached to each outer side of the first and second vertebræ is a hollow sphere of bone of equal size, between which, on the upper surface, the verte-

THE LOACH.

bræ are distinctly seen, but the union of the two spheres underneath hides the vertebræ when looking upwards from below. These bones are analogous to the scapulæ; to their outer surfaces the bones of the proximal extremity of the pectoral fins are articulated, and the fins moved by powerful muscles, which assist in producing the rapid motion observable in this little fish. Another peculiarity existing in the upper surface of the head is the want of union in the two parietal bones at the top, which occurs in the common guana; this peculiarity of the loach, it will be observed, is another instance of a relation in structure between the fishes and reptiles."

The loach seems to be extremely susceptible to atmospheric changes, and has often been kept as a "living barometer," its restlessness and constantly coming wriggling up from the bottom, where it usually lies quite still, denoting either storms or rain. This peculiarity some suppose to arise from its being what is called a ground-fish, having a low standard of respiration and a high degree of muscular irritability, which is rapidly acted upon by electrical changes in the atmosphere.

Some think the loach quite equal to the gudgeon in a gastronomic point of view. It is said that Frederick, King of Sweden, had the fish imported into Sweden for his own particular eating.

THE TENCH.

The TENCH (*Tinca vulgaris, Cyprinus tinca*). The scales are small, and deeply embedded in the thick skin; lateral line complete; dorsal fin short, its origin being opposite the ventral fin; a barbel at the angle of the mouth; pharyngeal teeth 4 or 5–5, cuneiform, slightly hooked at the end.

Izaak Walton says the tench "is observed to be a physician to other fishes, and has been so called by many that have been searchers into the nature of fish."

Thomas Weaver says:—

> "The tench, physician of the brook,
> In yon deep hole expects your hook,
> Which having first your pastime been,
> Serves then for meat or medicine."

And he also remarks that, in consequence of this medicinal power, the pike will not devour or hurt him.

Moses Browne, writing of the pike ("Piscatory Eclogues"), says:—

> "Yet, howsoe'er with raging famine pin'd,
> The tench he spares, a salutary kind.
> For when by wounds distrest, or sore disease,
> He courts the fish, medicinal, for ease.
> Close to his scales the head physician glides,
> And sweats a healing balsam from his sides."

This belief prevails to the present day. Buckland says: "I am not sure of the fact."

Day says: "It has been termed the 'Fishes' Physician;'" and Camden, in his "Britannica," observes that he has seen bellies of pike which have been rent open have their gaping wounds presently closed by the touch of the tench, and by their glutinous slime perfectly healed up.

This exemption, however, from the attacks of pike is very problematical. Day says: "In the 'Zoologist,' 1853, p. 4021, Mr. Slaney observed a pike which had seized a tench of about 3 lbs. weight, crosswise, and was unable to swallow it. A few days subsequently he saw another tench undergoing the same process, and afterwards saw a dead one which had evidently been injured by a pike. In the 'Zoologist,' 1867, is a record of a tench of 7 lbs. weight being taken out of the stomach of a pike of between 30 and 40 lbs., taken at Frogmore, Windsor."

Tench love still waters, and their haunts in rivers are chiefly among weeds, and in places well shaded with bushes and rushes. He is often found near sluices, and is rather fond of the mud.

Daniel, in his "Rural Sports," gives an account of a tench found in a piece of water at Thornville Royal, which had been confined in a hole under the roots of a tree for so long a time as to take the shape of the place of his confinement. His length from fork to eye was *two feet nine inches;* his circumference, almost to the tail, *was two feet three inches;* his weight, *eleven pounds nine ounces and a quarter.* The colour was also singular, his belly being that of a char, or a vermilion. This extraordinary fish, after being inspected by many gentlemen, was carefully put in a pond; but either from confinement, age, or bulk, it at first merely floated, and at last with difficulty swam gently away.

The discovery and account of this fish was, however, not to go unchallenged; and Daniel says :—

"To this account some sceptics have demurred, and have expressed their doubts in prose and verse, as follows :—

"'The *yellow-bellied* TENCH of *Thornville House,* in Yorkshire, which is *supposed* to have lain so many *centuries,* and lived under the *roots* of some ancient *trees* without *water,* is to be dressed at that celebrated mansion as soon as an instrument is procured in which a proper *kettle of fish* may be made of this amphibious *animal.* It is to be served up with *sauce piquant,* at a kind of *Arthur's Round Table,* to a select corps of the *Knights of the Long Bow.*'

"'THE TENCH OF THORNVILLE HOUSE.

A TRUE STORY !!!

"'OH, the marvellous,
   At *Thornville House,*
We read of feats in plenty;
   Where, with *long bow,*
   They hit, I trow,
Full nineteen shots in twenty!

"'Their fame to fix,
   'Midst other tricks,
In which they so delight, sir,
   These blades, pray know,
The *hatchet throw,*
Till it is out of sight, sir.

"'Of beast and bird
   Enough we've heard,
By *cracks* as loud as thunders;
   So now they dish
   A monster fish,
For those who *bite at wonders.*

"'The scullion wench
   Did catch a *Tench,*
Fatter than Berkshire hogs, sir;
   Which, pretty soul,
   Had made his hole,
Snug sheltered by some logs, sir.

"'Sans *water* he
  Had liv'd, d'ye see,
  Beneath those roots of wood, sir,
  And there, alack!
  Flat on his back,
  Had lain since Noah's flood, sir.

"'Now he's in stew,
  For public *goût*,
  And fed with lettuce-cosse, sir;
  In hopes the Town
  Will gulp him down,
  With good *humbugging* sauce, sir.'"

Daniel adds: "But notwithstanding the *squibs* and *witticisms* of incredulity, the account is authentic."

The tench is very tenacious of life; is a slow grower, rarely exceeding 5 lbs. The general colour of the body is a light greenish-olive, with a golden sheen, lightest along the whole line of the under surface. Fins dark brown; the lateral line commencing at the upper part of the gill-cover, descending, and then going straight to the tail.

Fin-rays: dorsal, 11–13; pectoral, 17; ventral, 9–10; anal, 9–10; caudal, 17.

The tench spawns about the middle or end of June. The female is generally accompanied by two males, and deposits her ova on the leaves of aquatic plants—often on the broad-leaved pond-weed, *Potomogeton natans*—hence in some counties this plant is called tench-weed.

The tench is said to hibernate in the mud in winter.

Günther says: "Only one species of 'tench' is known (*T. tinca*), found all over Europe in stagnant waters with soft bottom. The Golden Tench is only a variety of colour, an incipient albinism like the Gold-Fish and Id."

THE CRUCIAN CARP.

Cyprinoids *without barbels*:—

The CRUCIAN CARP (*Carassius vulgaris*) differs from the common carp in having no barbels; the pharyngeal teeth being in a single series, 4–4, and compressed, and the shape being more bream-like. In the common crucian carp the

THE BREAM.

body is almost quadrangular, and the dorsal profile much more elevated than the abdominal.

It has much the same habits as the common carp, perhaps somewhat more sluggish, keeping much at the bottom of the water, although it likes the sun's heat. It is extremely tenacious of life. Yarrell says: "I have known them recover and survive after having been kept out of water thirty hours."

The colour of the crucian carp is rather brilliant. The

THE CRUCIAN CARP.

back has a fine metallic lustre, the eye is golden-yellow, the belly white, and the fins are orange-red. This fish does not grow to anything like the size of the common carp; a pound fish is a large one. Yarrell records one brought to him from the Thames weighing 2 lbs. 11 oz.

The PRUSSIAN CARP (*Carassius gibeleo*), which by some has been considered another species, is only a variety of *C. vulgaris*, differing in the body being not so deep and more carp-like, the back not so elevated, and the tail being more forked. The common Gold-Fish is *Carassius auratus*.

THE BREAM.

The BREAM or CARP-BREAM (*Abramis brama*). This fish has not, like the carp or barbel, either strong bony rays or barbules; the body is deep and compressed; dorsal and abdominal line very convex; the base of the dorsal fin short, and placed behind the line of the ventrals; base of the anal very long; pharyngeal teeth in one row.

The fin-rays are: dorsal, 11–12; pectoral, 15–17; ventral, 10; anal, 27–31; caudal, 19.

The general colour yellowish-white, becoming browner by age, something like the carp, and hence its name Carp-Bream. The irides are golden-yellow; cheeks and gill-covers silvery-white. The pectoral and ventral fins are tinged with red; the dorsal, anal, and caudal tinged with brown.

Buckland says the bream "is quite an angler's fish." Izaak Walton's description of the bream stands good to the present day. "The bream," he says, "is a large and stately fish. He will breed both in rivers and ponds, but

THE BREAM.

loves best to live in ponds, and where, if he likes the water and air, he will grow not only to be very large, but as fat as a hog." He also remarks that "some say breams and roaches will mix their eggs and milt together; and so there is in many places a bastard breed of breams that never come to be either large or good, but very numerous."

Day says: "I have already remarked, p. 177" ("On the Roach"), "how hybrids between the roach, *Leuciscus rutilus*, and other *Cyprinidæ*, cannot occur in Ireland, because the roach of that country is *L. erythropthalmus*, or the rudd, which is termed a roach in the Emerald Isle. It is not, however, improbable that elsewhere the *L. rutilus* (the true roach) may also assist in the creation of hybrids with the bream."

THE BREAM.

The bream is gregarious, and found in many rivers, canals, ponds, lakes, broads, &c., the Norfolk broads being famous for this fish; it runs to a considerable size, from 7 lbs. to 11 lbs., bites freely, and affords good sport to those who like it. Thomas Weaver (first edition of Walton) says :—

> "The treacherous quill in this slow stream
> Betrays the hunger of a bream."

Bream spawn in May, and in the broads in the spawning time, or just before, the big ones roll about "like porpoises."

In the "Boke of St. Albans" the bream is described as a "noble fysshe and deynteous."

Another bream, called the BREAM-FLAT or WHITE BREAM (*Abramis blicca*), was first described by the Rev. Mr. Sheppard in 1824. Day says: "It does not appear to collect in large shoals, and is more commonly found with the rudd and roach than the true bream. It is lively, sportive, and tenacious of life. Sir John Lubbock observes that its mode of biting when angled for is singular, as it appears more prone to rise than to descend, and the float, consequently, instead of being drawn under water, is laid horizontally on the surface." But we have often observed precisely the same proceeding in the biting of the carp.

The head is larger and the fleshy portion of the tail deeper than in the carp-bream. The lateral line is not so low down, and the number of rays in the fins differ. The pharyngeal teeth are in two rows.

The fin-rays are: dorsal, 10; pectoral, 14; ventral, 9; anal, 22; caudal, 19.

The general colour of the sides is silvery bluish-white, without any of the yellow-golden lustre observable in the carp-bream; the eyes silvery-white, tinged with pink; the pectoral and ventral fins tinged with red.

The Genus LEUCISCUS is the fifth of the *Cyprinidæ*. It is widely distributed over both the Old and New World, and is known under the general name of "White Fish." The generic characters are :—Body covered with imbricate scales; dorsal fin commencing opposite, rarely behind, the ventrals; and the anal fin generally with from nine to eleven, rarely with eight (in small species), and still more rarely with fourteen, rays; the mouth without any structural peculiarities. There are no barbels; the pharyngeal teeth are conical, or compressed into a single or double series.

There are six British species.

THE ROACH.

The ROACH (*Leuciscus rutilus*, *Cyprinus rutilus*) is found in most of our rivers, but prefers the sluggish streams, frequenting the deeps by day and the shallows by night. It lives in companies. The colour of the upper part of the head and back is dusky green, with blue reflections,

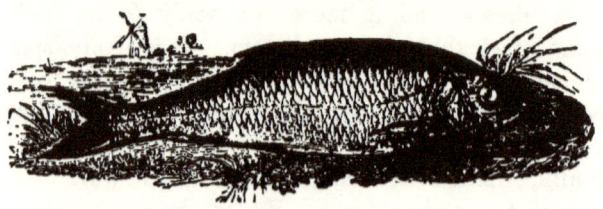

THE ROACH.

becoming lighter on the sides, and passing into silvery-white on the belly; the irides yellow; the cheeks and gill-covers silvery-white; the dorsal and caudal fins pale brown, tinged with red; pectoral fins orange-red; ventral and anal fins bright red.

Fin-rays: dorsal, 11–13; pectoral, 16–18; ventral, 9–10; anal, 12–13; caudal, 19.

The pharyngeal teeth are in a *single* row.

The roach rarely attains any great size. A fish of 2 lbs. is considered very large. Mr. Jesse ("Anglers' Rambles"), states that the largest roach he could hear of weighed 3 lbs. Pennant has recorded one of 5 lbs.

Roach spawn about the end of May or beginning of June, and when about to spawn immense shoals go "to

THE CHUB.

hill," as the fishermen call it, for the purpose of depositing their ova on the rushes and other aquatic plants. A great number of these ova are eaten by the water-shrimps and other aquatic insects. Buckland states that roach are subject to a curious disease, in which the scales turn jet black, and which is very fatal.

Moses Browne says: "The roach is called the river-sheep for his simplicity;" but the late Francis Francis writes: "No greater mistake can be made than to fancy the roach is so simple a fish."

Roach-fishing is an art of itself, requiring a quick eye, a quick strike, and very fine tackle. The Thames and Trent roach-fishers are great adepts, and, as Francis Francis says, "it is very pretty sport, requiring the exercise of much skill, patience, quickness of apprehension, and ingenuity, combined with a thorough knowledge of the habits of the fish."

The roach is not found in Ireland, its place being taken by the rudd.

THE CHUB.

The CHUB (*Leuciscus cephalus*), LARGE-HEADED DACE, SKELLY, CHEVIN.

The very obtuse snout, the large mouth, the wide and chubby head, and large scales distinguish this fish from

THE CHUB.

others of its kind. The lateral line descends by a gentle curve till it is even with the end of the pectoral fin-rays, then straight to the tail. The dorsal fin commences half-way between the point of the nose and end of fleshy part of the tail. Pharyngeal teeth, in a *double* row, pointed and hooked.

The colour of the top of head blackish-brown, with a streak of the same dark colour passing down the free edge of each gill-cover as far as the origin of the pectoral fin; the upper part of the back bluish-black; the edge of each scale darkest; sides bluish-white; silvery-white on belly; dorsal and caudal fins dusky; pectoral reddish-brown; ventral and anal reddish-white; irides golden-yellow; cheeks and gill-covers rich golden-yellow.

Fin-rays: dorsal, 10; pectoral, 16–17; ventral, 9–10; anal, 10–12; caudal, 19.

All who have read the "Complete Angler" know that the discourse on "Fishing" commences with an account of the "logger-headed chub," how to fish for him and how to dress him; ever since that time this fish has had its votaries, who are constantly recording their successes or their failures.

The chub is a very timid fish, but when hooked gives some sport. We have had occasionally a good tussle with a three or four pounder when fly-fishing for trout, their first rush being very powerful.

The chub spawns generally the latter end of April or early in May, depositing the ova on a gravelly bottom under weeds. After spawning it frequents the streams, soon regains its condition, and then returns to the deeper water. It is not found in Ireland.

THE RUDD.

The RUDD (*Leuciscus erythropthalmus*), RED-EYE, FIN-SCALE, SHALLOW.

The rudd is found in the Thames and in many other rivers of the United Kingdom. It takes the place of the roach in Ireland. Walton and others have considered the rudd to be a hybrid between the roach and the bream, but it appears to be a well-defined species. It is very plentiful in the Norfolk broads, and some attain to a weight of 2 lbs.; one is recorded of over 3 lbs. They bite very freely, both at the worm and at the fly. The late Mr. Manley says fishermen who want to make the special acquaintance of the rudd had better betake themselves to Slapton Lea and fly-fish for them on the sandy shallows in the summer months. The fish run up to 2 lbs., and are fairly sportive. He recommends that a little bit of white kid glove should be attached to the red palmer.

The colour of the rudd varies much. Buckland says: "In some shallow broads they are of a bright golden hue; in others, where the water is deep, their scales are like those of the roach or dace," as a general rule. The eyes are orange-red (hence the name Red-Eye), the cheeks and gill-covers golden-yellow; upper part of the back brown, tinged with green and blue; sides paler; abdomen light golden-yellow. The surface of the body has a brilliant reddish-golden hue, hence the name Rud. Moses Browne says:—

> "The Rud, a kind of roach, all tinged with gold;
> Strong, broad and thick, most lovely to behold."

The fins are of a carnation-red; the dorsal and caudal fins not so bright in colour.

Fin-rays: dorsal, 10; pectoral, 15; ventral, 9; anal, 13; caudal, 19. The Irish specimens of the rudd from Lough Neagh, Yarrell states, have one ray more in the dorsal and anal fins.

The rudd spawns about the end of April, depositing its eggs amongst the leaves of aquatic plants.

THE DACE.

The DACE (*Leuciscus vulgaris*), called also the DARE or DART, is a much more local fish than the roach. Like the chub, it will rise at the fly at times very freely. We have had many a good afternoon's sport, years gone by, at low water on the shallows below Teddington weir, with a black palmer. The muzzle is more pointed, the mouth rather larger and more deeply cut, than in the roach, and the eyes not so large. The back, also, is but slightly elevated, and the form of the body more elongated and more elegantly shaped; the scales are much smaller than those of the roach, and the dorsal fin commences rather more posteriorly. The colour of the upper part of the head and back dusky blue, paler on the sides; belly white; eyes straw-colour; gill-covers silvery-white; dorsal and caudal fins pale-brown; pectoral, ventral, and anal almost white, tinged with red.

Fine-rays: dorsal, 9–10; pectoral, 15–16; ventral, 9–10; anal, 10–11; caudal, 19–20.

The dace spawns at the end of May and beginning of June, depositing its eggs at the roots of aquatic plants, under stones, and on the gravelly beds of rivers. The best season for fishing for dace is the end of July onwards.

THE DACE.

Buckland says that "at certain times of the year they assemble in vast numbers at the lower side of the Thames navigation weirs, on the road up-stream. We generally get leave from the Thames Conservators to open one of the hatches and let them pass up."

Dace are not found in Scotland or Ireland.

Whether the AZURINE (*Leuciscus cæruleus*) is a distinct species or a variety of the dace is a question. It was first added to the British list by Yarrell, and is found only in a few localities; its local name is the Blue Roach. Day ("The Fishes of Great Britain and Ireland," vol. ii. p. 184), in describing the varieties of the rudd, says: "The azurine was added to the list of British fishes, some fish of this variety having been obtained from Knowsley.

It is less bright, its abdomen silvery, and its fins white." Mr. Pennell found in some ponds near Romford, in Essex, a lemon or yellow coloured variety of the rudd.

THE BLEAK.

The BLEAK has been removed into a genus of itself, under the name of *Alburnus*. The scientific name was formerly *Leuciscus alburnus*. It is now *Alburnus lucidus*.

THE BLEAK.

This well-known fish, so abundant in many rivers, is much used as a bait in spinning, for either trout or pike. It is generally from five to seven inches in length; dace-like in form, but immediately distinguishable from the backward position of the dorsal fin and the greater length of the base of the anal. The colour of the back is a bluish-silver, tinged with green; very silvery sides; eyes silvery, tinged with yellow; all the fins whitish.

Fin-rays are: dorsal, 10–11; pectoral, 16–17; ventral, 9–10; anal, 19–23; caudal, 19.

The bleak spawns in May and June.

The name Bleak is supposed to be derived from the Saxon word *Bleage*, or German *Blick*, *Blicken* to shine. It is also called Blick, signifying to bleach; it has likewise a number of local names, as Blaze, Blay, Willow-Blade, Ablet, Fresh-Water Sprat. In France it is called *L'Ablette Commune*. It is not found in Scotland or Ireland.

The scales of the bleak were formerly used for making false pearls, a discovery of a Monsieur Jaquin, who kept his secret for a very long time. The scales of two hundred and fifty of these fish will not weigh more than an ounce, and these will not yield more than a fourth of that weight of pearly powder applicable to the preparation of beads, so that 16,000 fish were required in order to obtain only one pound of the essence of pearls. The manufacture of false pearls from bleak-scales has greatly diminished, their place being taken by the silvery pigment found in the air-bladder of the Argentine (*Argentina sphyrœna*), a Salmonoid found on the shores of Norway, on both the east and west coasts of Scotland, and elsewhere.

THE MINNOW.

The MINNOW, MINIM, or PINK (*Leuciscus phoxinus*) is one of the smallest, but as regards colour one of the most beautiful, of the British *Cyprinidœ*. It seldom exceeds from three to four inches in length, and is found in many rivers, brooks, and canals. The figure given is one exceptionally large—four and a half inches—sent to the editor of the *Fishing Gazette*. The Minnow is very common in England; rare in Scotland and Ireland.

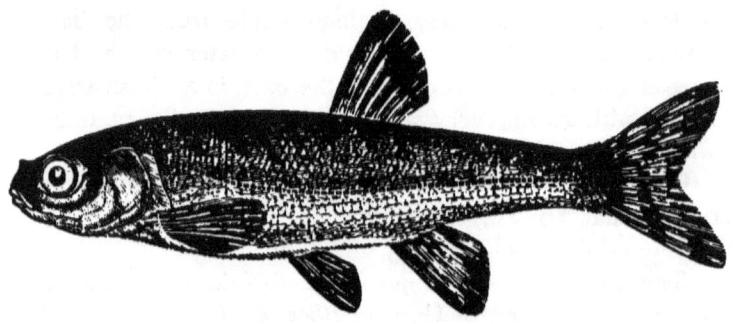

THE MINNOW.

The name is supposed to be derived from the Latin word *minimus*, little. It has several provincial names— Pink, from its colour; Bennick, Minim, Peer, Jack Barnell, Meaker, Mengy, Mennard, Shabrid. In France it is called *La Vairon*.

The minnow is a most prolific breeder, spawning about the end of May, and then frequenting the ditches and small watercourses running into the main stream. When about to spawn the head is usually covered with small white tubercles; it is also at this time, as Buckland says, "the gentlemen wear green and red coats."

Minnows are voracious feeders, and it is doubtful if their presence in trout-streams is beneficial when very numerous. No doubt they are food for the big fish, but do not they devour an enormous quantity of food of which the smaller trout ought to be the sole recipients?

How well Keats describes their habits!—

> "O'er their pebbly beds,
> Where swarms of minnows show their little heads,
> Staying their wavy bodies 'gainst the streams,
> To taste the luxury of the sunny beams,
> Tempered with coolness. How they ever wrestle
> With their own sweet delight, and ever nestle
> Their silver bellies on the pebbly sand!
> If you but scantily hold out your hand,
> That very instant not one will remain;
> But turn your eye, and there they are again."

The colour on the top of the head and back is darkish-olive, mottled, and of a silvery-pink colour on the sides; the abdomen white, with a beautiful rosy or pink tint, varying in intensity; the irides and gill-covers silvery-white; dorsal fin pale-brown; pectoral, ventral, and anal fins much lighter in colour; the caudals light-brown, with one dark-brown spot at the base of the rays.

Fin-rays: dorsal, 9; pectoral, 16; ventral, 8; anal, 9; caudals 19.

THE SHAD.

The SHAD or TWAITE SHAD (*Alosa finta*), one of the *Clupeidæ*, is an anadromous fish, entering the fresh-waters about May and ascending the rivers for spawning purposes to a considerable distance. It was formerly abundant in the Thames, and one of its favourite haunts were the reaches between Putney Bridge and Hammersmith. There are very few, if any, to be found now in the Thames, but in the Severn they are more common, and generally follow the advent of the larger shad, *Alosa communis*, the Allice Shad. Buckland says the twaite is distinguished from the other native species by its having distinct teeth in both jaws, and by a row of rather large dusky spots along each side of the body; while in the allice shad there is only one spot, and that close to the side of the upper part of

THE TWAITE SHAD.

the gill-cover. The allice is a much larger fish, found in the Wye and Severn. Day says that the twaite shad is very capricious of taking a bait in the Severn, being occasionally hooked at almost any cast, at another time playing around the bait, but refusing to touch it. When hooked it affords some play.

CHAPTER XIV.

THE SALMONIDÆ.

DR. GÜNTHER ("Study of Fishes") places the *Salmonidæ* as the fifteenth family of the order PHYSOSTOMI.

The first genus of this family comprise the Salmon, the Trout (*Salmones*), and the Char (*Salvelinus*); the latter being divided from the former by having no teeth on the body of the vomer, but only on the head of that bone.

The British representatives of the first group are—

    The Salmon (*Salmo salar*).
    The Sea-Trout (*Salmo trutta*), with its varieties or species, as—
        The Sewin (*S. cambricus*).
        The Bull-Trout (*S. eriox*).
        The White Trout (*S. albus*).
    The Fresh-Water Trout (*Salmo fario*), with its varieties or species, as—
        The Loch Leven Trout (*S. Levenensis*).
        The Crass Puil Trout.
        The Estuary Trout (*S. estuarius*).
        The Orkney Trout (*S. orcadensis*).
        The Cornish Trout (*S. cornubiensis*).
        The Great Lake Trout (*S. ferox*).
        The Gillaroo Trout (*S. stomachicus*).

The British representative of the second group, *Salvelinus*, is—

    The Char (*Salmo alpinus*), with its varieties or species, as—
    Torgoch (*S. cambricus*).
    Loch Killin Char (*S. Killinensis*).
    Gray's Char (*S. grayii*).
    Coles' Char (*S. colei*).

The second genus, *Thymallus*, has only one British representative—

    The Grayling (*Thymallus vulgaris*).

The third genus, *Coregonus*, placed by some as the second, has three representatives—

    The Guyniad, Schelly, or Powan (*Coregonus clupeoidis*).
    The Vendace or Vendis (*Coregonus vandesii*).
    The Pollan (*Coregonus pollan*).

The fourth genus, *Osmerus*, only one representative—

    The Smelt (*Osmerus eperlanus*).

Much difference of opinion exists as to whether many of the British forms of the genus *Salmo* are to be considered as constituting species or varieties.

The term species in natural history is employed to designate groups inferior to genera, but superior to varieties. Formerly, it was supposed the species were unchanging throughout the longest successions of generations, and would be defined as individual plants, animals, &c., agreeing in their appearance and composition, their similarity giving rise to the establishment of species; and individuals or species differing in circumstances arising from accident in plants or animals, from soil or climate, were termed varieties.

About the end of the seventeenth century John Ray limited the term species in its natural history sense. His specific characters rested more especially on constant resemblance in outward form, but also on close resemblance of offspring to parent. At the same time he recognised variability.

Linnæus's aphorism, "*Tot sunt diversæ, quot diversæ formæ ab initio sunt creatæ*," notwithstanding Buffon's objection, was in a great measure recognised by Cuvier, and held in general acceptance until Darwin subverted it in 1859 on the publication of the "Origin of Species."

Darwin maintained the variability of species, the variation continually taking place according to the external condition to which plants and animals are exposed. He thinks it difficult to distinguish varieties from species, and refers to the well-known changes which are produced by cultivation and domestication. He instances the selection man makes in order to produce new breeds and varieties, and considers that such selection often takes place in nature in the struggle for life which all plants and animals must undergo.

Professor Huxley advocates the hypothesis which supposes the species living at any time to be the result of the gradual modification of pre-existing species. Day ("British and Irish Salmonidæ") says: "We must have a permanence of variation from the original form to indicate a distinct species." He goes on to say, "that owing to too great importance having been given to inconstant variation, the number of species among this family has been unduly augmented, and varieties have been accorded specific rank, while every little variety of form, colour, or structure has also been reckoned as possibly demonstrating hybridity."

The author of the article "Species," in the "Encyclopædia Britannica," says: "The rash generalisation, that distinct species are to be recognised by their incapacity for the production of fertile hybrids, has been overthrown, while closer study has cleared away the notion of the equal definitiveness of specific forms.

"The want of any absolute standard of specific difference is largely made up by practical experience and common sense, and the evolutionary systematists are less in danger than were their predecessors of either exaggerating or understating the importance of mere varieties."

The *Salmonidæ* are described by Dr. Günther as follows:—"Body generally covered with scales, head naked, barbels none. Margin of the upper jaw formed by the intermaxillaries mesially, and by the maxillaries laterally; belly rounded; a small adipose fin behind the dorsal; pyloric appendages generally numerous, rarely absent; air-bladder large, simple; pseudo branchiæ present. The ova fall into the cavity of the abdomen before exclusion.

Salmo.—Body covered with small scales; cleft of the mouth wide, the maxillary extending below or beyond the eye; dentition well developed; conical teeth in the jaw-bones, on the vomer and palatines, and on the tongue; none on the pterygoid bones; anal fin short, with less than fourteen rays; pyloric appendages numerous; ova large; young specimens with dark cross bands (parr marks)."

"The *Salmo salar*," says Günther, "can be generally recognised, but there are instances in which the identification of specimens is doubtful, and in which the following characters, besides others, will be of great assistance. The tail is covered with relatively large scales, there being constantly eleven, or sometimes twelve, in a transverse series, running from behind the adipose fin forwards to the lateral line, whilst there are from thirteen to fifteen in the different kinds of sea-trout and river-trout. The number of pyloric appendages is great, generally between sixty and seventy, more rarely falling to fifty-three, or rising to seventy-seven. The body of the vomer is armed with a single series of small teeth, which at an early age are gradually lost from behind towards the front, so that half-grown and old individuals have only a few (one to four) left."[1]

To this description may be added another very marked distinction. As a rule, salmon rarely have any spots below the lateral line, except one or two by the gill-covers, whereas in the trout—either sea, bull, or river—the spots extend considerably below that line.

THE SALMON.

The SALMON (*Salmo salar*), after leaving the egg, goes through four distinct stages of its existence, viz., the Parr, the Smolt, the Grilse, and the Salmon proper.

It is of very great importance to all who fish rivers or lakes where salmon have bred, to be able to distinguish a salmon-parr from a trout-parr. The former it is always illegal to kill, whilst the latter can be taken without in-

---

[1] A full account of the salmon and its allies is to be found in Dr. Day's "British and Irish Salmonidæ," a book of great value, both for instruction and for reference.

THE SALMON.

fringing any of the laws laid down for the purpose of preserving the salmon and trout. How, then, are we able to distinguish the one from the other? Dr. Day ("British and Irish Salmonidæ") asks the question, "What is a parr?" and says: "The young of the salmon in Acts of Parliament were formerly designated as fry and smolts; while of late years the term parr has been commonly used, which has been said to be calculated to mislead, because there are salmon-parr and trout-parr."

Yarrell ("History of British Fishes," first edition, 1836) makes the parr or samlet a distinct species; and at that time this was the general opinion. He also says: "It is this similarity in the marking and appearance of the fry which has caused the difficulty in distinguishing the

THE SALMON.

various species when so young; and experimenters, believing they had marked young parr only, have been surprised to find some of these marked fish return as grilse, young bull-trout or whitling, salmon-trout, river-trout, and true parr;" and further on states "that this little fish, one of the smallest of the British *Salmonidæ*, has given rise to more discussion than any other species of the genus." But in the second edition, 1841, he says: "In order to prevent any misconception of the terms employed, I shall speak of the young salmon of the first year as a *pink;* on its second year, until it goes to the sea, as a *smolt;* in the autumn of the second year, as a salmon-peal or grilse; and afterwards as an adult salmon." In a subsequent edition he gives up the parr as a distinct species.

Parnell ("Fishes of the District of the Forth, 1638) considered the parr, or *Salmo samulus*, as a distinct species; and he describes the two (parr and young salmon), both taken from the same river in the month of May. He says : "The form of the young salmon is long and narrow; the snout pointed; the caudal fin acutely forked. The body of the trout-parr is thick and clumsy; the snout broad and blunt; the caudal fin much less forked.

"The operculum of the salmon-parr is beautifully rounded at its posterior margin, and the basal line of union with the sub-operculum much curved. In the trout-parr this part is nearly straight with the line of union.

"In the salmon-parr the maxillary is short and narrow; in the trout-parr it is broader and longer. The pectoral, dorsal, and caudal fin in the salmon-parr are blue, in the true parr dusky. The flesh of the salmon-parr is delicate and pinkish, the bones soft; the flesh of the trout-parr is white and firm, the bones stout and hard."

He goes on to say : "It is generally supposed that those small fish, from four to five inches in length, which are found so plentiful in many rivers during the autumn months, and which are marked on the sides with from ten to eleven transverse dusky bands, and a black spot on each gill-cover, are either all parrs or the young of a salmon; but from a minute examination of several hundreds of these fish taken in various rivers in England and Scotland, I am induced to consider them not all of one species, but the young of various species or varieties of migrating trout, in company with the young of the salmon, with the *Salmo samulus* or parr, and with different varieties of the fresh-water trout, all of which have received the names of Heppers, Brandlings, Samlets, Fingerlings, Gravellings, Lasprings, Skirlings, and Sparlings."

Thompson, in his "Natural History of Ireland, 1856, says: "The three most striking characters of the parr, in contradistinction to the common trout, are its tail being more forked, its having only two or three spots on the opercula, and its want of dark-coloured spots beneath the lateral line. The pectoral fin of the parr is larger, and the hinder-part of the operculum less angular, than in that of the common trout."

In the Tweed Salmon Reports, 1867, Mr. Paxton, Superintendent of the Tweed Fisheries, says: "The salmon-smolt is easily distinguished from the trout-smolt or orange-fin, and from the common river-trout of the same size, by its more slender and tapering body, silvery scales, and black fins. The orange-fin is not so easily distinguished from the river-trout of the same size; although its white belly, bright orange fins, and fewer and smaller spots cause it to be easily detected by an experienced eye. During the first year of their life the young of three species of *Salmonidæ* belonging to the Tweed are scarcely to be distinguished, and are indiscriminately called *parr*. They have dark bars across their sides, which are owing to different shades of colour in the true skin; for, on removing the scales from the smolt, those parr-marks, as they are called, make their appearance." And in answer to Question 49, he says: "When three to six inches in length, the parr-marks disappear, and the salmon-smolt gets a coating of silvery scales, its fins become black; the trout-smolt has its pectoral fins *bright orange*."

Mr. Alexander Mitchell, same Report, says: "The salmon and sea-trout fry are not, in my opinion, distinguishable until they respectively assume the smolt state."

Mr. Peter Marshall, says "that the parr of the trout has the dead (adipose) fin orange. The rudder-fin is white at the bottom and yellow at the top. They have not so many parr-marks as the parr."

Mr. Russel ("The Salmon," 1864) says, "Every schoolboy on the banks of the Tweed knows at a glance the difference between the smolt of the salmon and of the bull-trout, the black-fin and the orange-fin."

Dr. Günther ("Evidence on Trial," June 4, 1872) says: "There is a distinction between the young *Salmo salar* (the salmon) and a member of the *Farios* or trout. In the parr of the former I have counted as many as nine or ten crossbars, and in the latter only six or seven."

Mr. Cholmondeley Pennell ("Angler Naturalist") says:

"According to Sir W. Jardine, the fry of the common trout, *S. fario*, may always be distinguished from that of either of the three migratory species, by its having the extremity of the second dorsal or adipose fin fringed with orange—a mark easily identified."

Mr. Shaw (keeper to the Duke of Buccleuch) was really the first person who demonstrated the identity of the salmon and parr. He gives the following representations of the salmon-parr, sea-trout parr, and river-trout parr :—

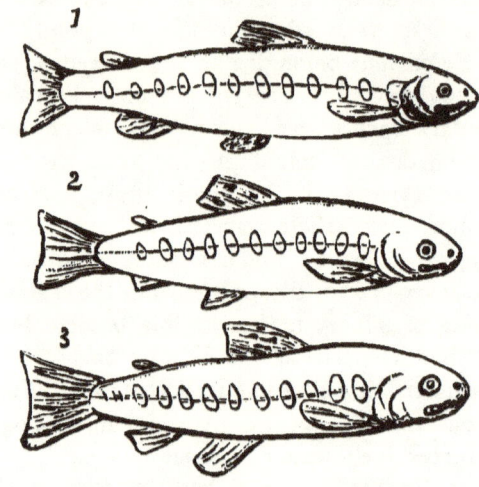

1. The Salmon-Parr (*S. salar*).
2. The Sea-Trout-Parr (*S. trutta*).
3. The Trout-Parr (*S. fario*).

Putting aside the internal evidences of distinction between the different kinds of parr, that which is most essential to the angler is the external appearance by which he would be able at once to recognise whether he was doing a legal or an illegal act by basketing the fish he has caught. As far as we can judge from studying the evidences of various writers, it appears that :—

In the salmon-parr the body is long and graceful; the head and snout longer; the parr-marks (transverse dark bands) very distinct, and separated by broad intervals; the pectoral fins narrow, and with the ventral and anal, of a dusky hue; the tail much forked.

In the migratory trout the body is thick and short; the head and snout more rounded; the dorsal fin often spotted; the pectoral broad, and with the ventral and anal orange-colours; the adipose fin tinged at the end a light orange; the tail but little forked.

In the river or brook trout the body is long and not so shapely; the head short; the snout very obtuse; the eye large; the dorsal fin spotted; the adipose fin with a scarlet red tip; the tail square, and but little forked. To the inexperienced, and even to some of the experts, the absolute certainty of the distinction is often a matter of difficulty.

Mr. Willis Bund (in "Salmon Problems") writes:—" I recollect once hearing a man tried for taking samlets. He was very well defended; and the solicitor asked the witness, who swore positively to the fish being a samlet, if he could always identify a samlet from a trout. To my horror, the witness said 'Yes.' 'Are you sure?' 'Certain; I could not mistake.' Four bottles were then produced, marked A, B, C, and D. In these were fish preserved in spirits. The witness was asked, 'What was A?'—'Samlet.' 'B?' —'Samlet.' 'C?'—'Trout.' 'D?'—'Samlet;' and he stuck to it, and pointed out the distinctions to the Court. For the defence, a witness was called who had bred the fish and placed them in the bottles. They were all four the same, and hybrids. I am bound to say that I was very nearly falling into the trap myself, and should certainly have said the fish were not the same, only I considered D was the trout, and the other three samlets. It is only, however, very rarely that these cases occur; usually any person who has had any experience can tell a samlet without any doubt."

According to the Salmon Act of 1861, the names given or applied to the young of the salmon (*S. salar*) are: fry, samlet, smolt, smelt, skirling or skarling, parr, spawn, pink, last spring, kepper, last brood gravelling, shed, scad, blue fin, black tip, fingerling, brandling, brondling, or by any other name, local or otherwise—evidently a determination to take in every provincial name. Surely a more simple mode of defining "What is a young salmon?" might have been arrived at.

Mr. George Rooper (in the last edition of "The Thames and Tweed") says: "The fry in its early stage is barred, and improperly called a parr. The true parr, *Salmo salmulus*, of Linnæus, Yarrell, and many other authors, differs materially from the salmon-parr, but they are sufficiently like for confusion; and much confusion and much bitter discussion have ensued from the two fish being called by the same name." Mr. Rooper believes in the existence of the *Salmo salmulus* as a distinct species, and he points out the distinction of the two in *Land and Water*, in answer to Mr. Walsh:—" The transverse bars on *Salmo salmulus* are more numerous, more strongly marked, longer and narrower than the parr of the salmon. The colour of the eye is totally different, though I cannot describe the difference. The gill-covers of the *Salmo salmulus* are invariably spotted, those of the smolt-parr are never. The pectoral fins are longer, larger, and stronger; as befits the fish that 'rides the rack,' and habitually frequents the streams the infant salmon could not for its life ascend." He also says "that *Salmo salmulus* has the red spot on the tip of the adipose fin, proving him to be a *Fario*. No migratory parr has this mark."

The young salmon remains in the parr state from about two months old to about fifteen months, when it then commences to put on its travelling dress—those delicate silvery scales—preparatory to its journey to the sea, which is usually completed about the end of April. It then becomes the smolt; and, as the old rhymes say—

"The last spring floods that happen in May
Carry the salmon fry down to the sea."

"The floods in May
Take the smolts away."

Dr. Day says that the silvery colour is not owing to their acquiring an additional coating of scales, as has been

THE SALMON.

asserted, but is due to deposition of silvery pigment on the under-surface of the scales and opercles. Dr. Günther considers that a new coat of scales overlays the parr-marks.

Usually the smolts do not go direct to the sea, but remain for a short time in the brackish water, to acclimatise themselves, as it were, to the change, and then go farther into the salt water—whether in the estuaries only or direct to the sea is not precisely ascertained; but they go where they find a great abundance of the food necessary for their rapid growth, for no such rapid growth in any other department of the animal kingdom is known as that of the salmon as it passes from the smolt to the grilse state of its existence. Marked smolts which have gone down to the sea in the spring some six or eight inches in length, and weighing from 5 to 7 oz., have returned in the autumn from 2 to 8 lbs. in weight, and even heavier.

Mr. Willis Bund (*loc. cit.*) says: "Wherever it is that smolts go, the next question is, How long do they stay? It is usually answered, a year, or even less; that is, that the smolts which go down one spring will return as grilse certainly the next year, possibly before. There is really no evidence to show this is correct, and judging from the previous growth of the fish it seems very likely it is not. At the lowest computation, it has taken a smolt a year (more likely two) to reach the size of six inches, and that under favourable circumstances as to food and water and climate. Why should he in the second year of his life grow three or four times as much as in his first year, and increase from ounces to pounds? It may be so, but it is so contrary to what one would expect that one requires some strong evidence to make us feel sure it is so."

Unless we disbelieve the Stormontfeld and other experiments, the evidence is strong enough to prove that at least a good many smolts of the spring return as grilse in the autumn of the same year.

Mr. Young of Invershin, in April and May 1837, marked a great number of descending smolts by making a peculiar perforation in the caudal fin with a pair of small nipping-irons constructed for the purpose. In the months of June and July (mark how short the time) he caught a considerable number on their return to the rivers, all in the grilse state, and varying from 3 to 8 lbs., according to length of their sojourn in the sea. Again, in 1842, similar experiments led to similar results.

In the Stormontfeld experiments, Mr. Brown, in May 1855, marked 1300 smolts by cutting off the adipose fin. Twenty-two of these were recaptured as grilse the same summer: the first on July 7, weighing 3 lbs.; the last on August 14, weighing 8 lbs.; the remaining varying between these two weights. On the other hand, some remain in the sea much longer. The Duke of Roxburghe, in May 1855, had a smolt marked by the insertion of a peculiar-shaped wire through the gills; it was retaken on July 21, 1856, as a grilse, weighing $6\frac{1}{2}$ lbs. The late Frank Buckland was of opinion that some of the smolts stay one, or even two, years in the sea. "Grilse," he says, "have even been captured of the extraordinary weight of 14 lbs.," &c. "Such a striking augmentation of size has in all probability resulted from the operation of various causes, a longer stay than usual on the feeding-grounds, or a richer or more abundant supply of food, &c." We ourselves, in August 1855, caught a grilse in the Lady Saltoun Pool, on the Ness, weighing 12 lbs., and one was taken in 1867 in the Tweed of $14\frac{1}{2}$ lbs.

For further information on this interesting subject, we would refer our readers to the very copious notes in Day's "British and Irish Salmonidæ."

The following letter appeared in *Land and Water*, March 10, 1888, doubting the great increase in size and weight of the smolt to the grilse in a few months, at least in Devonshire; but are not the fish termed "peel" by Mr. Pike sea-trout and not salmon? and is not the salmon-peel, salmon-trout, the same as the sea-trout, *S. trutta?*—

"Sir,—What is a grilse? at what age and size is it

THE SALMON.

first termed a grilse? and what is the fish termed at the intermediate stage between a smolt and a grilse?

"These questions suggest themselves to me by the statements in Seeley's book of the 'Fresh-Water Fishes of Europe.' Speaking of *Salmo salar*, at p. 274, he says the young 'parr' are hatched in February or March; by the middle of May they are an inch long. In a year they increase to $3\frac{1}{2}$ inches, and in two years they are $6\frac{1}{2}$ inches long, when they become smolts, and flit from their native river to the sea. 'They revel in the sea, and when they come back in a few months it is as grilse, with boisterous energy.'

"Now, this description does not accord with my experience, and I have now been eighteen years connected with the Board of Conservators for the Dart District.

"I have never heard the term grilse applied to young, fresh-run salmon under $2\frac{1}{2}$ lbs. in weight, and I certainly cannot believe the young smolts can increase from about 2 oz. in weight to $2\frac{1}{2}$ lbs. in the one short season of their first going to sea—'a few months,' not more than three, as grilse are taken in August.

"My experience is, that the young are hatched from January to March, and some even much later. The great majority of these put on the silvery coat when they are about fifteen months old, and go down to the sea from March to June, as the freshets occur to help them. These little ones are then only about 2 oz. in weight. They return again from the sea in July and August, but certainly not as grilse. We then term them here peel, and they vary in weight from $\frac{1}{2}$ lb. to $1\frac{1}{2}$ lb., the greater number being about 1 lb. each. The following season they increase to about 3 lbs. to 6 lbs. each, and are then what is usually termed grilse, or in this district harvest salmon.

"I believe that the young salmon return to all rivers in the stage of peel, but that they are then mixed up with other breeds of the genus salmon, and all are termed alike 'sea-trout;' they are, however, easily distinguished from the sea or bull trout, which is here termed a truff.

"It would be interesting if some of your correspondents would state their experience as to the progressive growth of these fish from smolts to salmon.

"Anthony Pike, Sec.,
"Dart Board of Conservators."

From experiments lately made by Mr. Walter C. Archer in the Sands river, Norway (see the *Field*, January 5, 1889) it is found that the increase of weight in a salmon recaught after an interval of from seven to ten months is much greater than when the interval has been twelve months.

A salmon, marked in October 1885 weighing 18 lbs., was recaught July 1887 weighing $30\frac{1}{2}$ lbs.

Mr. Archer thinks it possible for a salmon to grow to 30 lbs. weight from the grilse state in four and a half years.

The grilse then having gone down to the salt water in the spring as a kelt, whenever he does return to his native river it is as a salmon; the size and weight depending on the food and other causes, of which, at the present time, we have but slight knowledge. As the lordly salmon, the king of fresh-water fish, we now meet him.

There is no British fish whose life-history has caused so much discussion, so much difference of opinion, so many controversies, or is so much prized—both for its sporting attributes and its edible qualities—as the salmon.

That part of its life which is passed in the salt water is still involved in doubt. It appears for the time being to vanish, as it were; and whether it goes far out to sea or remains in the estuaries, or whether it goes north or south, we know not. When it returns to what may be considered its native element—the fresh water—we are able to trace its progress to the upper waters, and to note its history during its sojourn, until it again goes to the sea to recruit the powers it has lost during the process of reproducing its species.

THE SALMON.

Without doubt, certain facts have been ascertained, particularly as to different phases this fish goes through in early life, from the parr to the smolt and grilse state, and so to the mature salmon, analogous to the baby, boy, youth, and man state in the human subject. But it is only necessary to read Mr. Willis Bund's "Salmon Problems" to find out how very little we know of its life.

The salmon, both in its smolt and its mature state, has many enemies lying in wait for it in the estuaries of the rivers—viz., the seals, the porpoises, the grampus, the coalfish, &c., whilst the otter watches for it in the fresh water. Even some birds will attack this fish. The great black-backed gull, *Larus marinus*, is known by the name of the salmon-gull amongst the fishermen of the lower part of the Severn, where, should a salmon be stranded on the sand, as sometimes is the case from the rapid fall of the water, these birds at once attack it, pick out its eyes, and soon tear it to pieces with their strong and powerful bills. Even the heron—it must be out of wantonness—will at times attack a salmon. We found a dead salmon, 19 lbs. in weight, evidently killed by a stroke of the sharp-pointed bill of this bird, which had pierced the brain.

Having escaped all these enemies—man included—the salmon arrives at the upper waters for the purpose of spawning. The male, or cock fish, when that period arrives, some time between October and December, entirely changes his appearance; he becomes what the fishermen call a red fish, and when caught the salmon-fisher exclaims, "What an ugly brute!" But is he so? He may be ugly in the fisherman's sense of beauty in a fish, but he is not so to his lady. Why has he put on this change from the blue silvery-steel appearance of his back and the transparent whiteness of his sides, with a sheen of iridescence, to the olive-greenish hue of the upper part, and the red-scarlet colour of the sides and yellowish-red belly, with the beautiful-shaped head of his early days transformed into prolonged jaws and a great beak? Because he is able evidently to captivate his love by his beautiful clothes according to her ideas. *Chacun à son goût.*

In this state the fish in Scottish parlance is a kipper. This term is derived from the hook or kype of the male. Mr. Brander, *Field*, October 1886, remarks: "A kip-nosed man, in Scotch, means a man with a turned-up pug nose." Reference to a Scotch Dictionary will show that anything turned up at the corners is said to be kippered.

A female full of spawn is a baggit, and, as we all know, after spawning and milting both go by the name of kelt. We were very much taken aback the first time we came in contact, when fishing in the Usk some years ago, with a kelt, having got somehow or other a different idea, thinking it must be something like a trout out of season, black and discoloured, instead of finding it with silvery sides and a bluish-green back instead of grey-blue silver. We recollect getting hold of one fine fellow, who gave us grand sport, taking twenty minutes to get him in; he weighed 18 lbs., a male fish, bright and silvery, still only a kelt, but a mended one. But where was his beak? There were the curved jaws, but no hook or kype.

Very much has been written concerning this appendage to the male salmon and its uses. At one time it was supposed that the kype was for the purpose of forming the spawning-bed for the female, which it could not possibly do unless the fish turned on its back or opened its mouth wider than a hippopotamus. It is now well known that the female scoops out the cavity in the gravel chiefly by means of her tail and lower part of the body, and that the hook of the male is for defensive purposes.

In the *Fishing Gazette*, March 1886, a correspondent, under the name of "Piscator," relates how he saw—"on a spawning-bed"—"six or seven male fish surrounding a female. There was one fellow who appeared to be king; whenever one attempted to come too near the female, he made a rush at him, seized him, and shook him as a terrier-dog would shake a rat. This game was carried on for a considerable time, until every one was more or less torn to such an extent that in a few days several of them were found lying dead on the bank, and I had no trouble in identifying them as those engaged in the combat."

THE SALMON.

Shaw ("Development and Growth of Salmon Fry") mentions "that he saw two males keeping up an incessant combat during a whole day for the possession of the female, and in the course of their struggles frequently drove each other almost ashore, and were repeatedly on the surface, displaying their dorsal fins and lashing the water with their tails." No wonder, then, that so many kippers die.

The beak of kype after the spawning season is over, disappears. But how? Day says: "The knob appears to be entirely composed of connective tissue, so cannot fall off, but may be more or less absorbed, as it doubtless is after the breeding season." In a note he gives Professor Gadow's opinion, "who was so good as to make sections of one and stain it with carmine." Professor Gadow says: "The hooks consist entirely of fibrous connective tissue without any trace of cartilaginous cells in it, the whole thing being surrounded by an epiderm. Therefore, the hook cannot be looked upon either as an out-growth of the bones of the lower jaw or as a sort of horny excrescence, like horns, nails, or pads, such as toads possess on the palms of their hands, but as a periodical out-growth of the cutaneous connective tissue which surrounds the body, being situated between the epiderm and the bone, without, however, having any relation to the periosteum. This cutaneous connective tissue, explains why and how the hook can again be absorbed, or rather reabsorbed, after the season is over; certainly it cannot be shed."

Year after year we have examined the jaws of the large male salmon which are exposed for sale at the fishmongers' shops. In some there is but a sign of the new beak, but *always* a circular thickening of the tissue covering the jaw; in others merely the rudiments of the kype; and again in others the hook is a quarter to half an inch high. In all the large ones the jaws are arched, particularly the upper one, and the hole in the inside of the upper jaw—where the point of the former kype rested—intact.

A correspondent in the *Field* newspaper, however,

states that: "When the fish has reached a certain stage in the kelt state, the hook gradually loosens at what seems on examination to be a kind of joint, just where the point of the nose should be in the fish. A slight tap when it has arrived at this state, or slight pressure on the gravel, will dislodge it. I have dislodged many."

It has been noticed that at times female salmon have developed a kype. Why should they not? The same thing occurs in other divisions of the animal kingdom, either from age, or injury, or absence of the reproductory organs. The female pheasant has often been found to put on some of the plumage of the male, and other birds the same. Hinds have been known to have horns; women, beards and whiskers.

Mr. Harvie Brown (*Zoologist*, May 1886) says that he and his friend, when fishing a river in Ross-shire, killed two fish having a horny projection, both of them having well-developed ova in them.

It is now, we believe, fully proved that there is an autumn migration of smolts; if so, do these fish return in the spring as grilse, or do they remain in the sea over the year?

The salmon smolt on returning to the fresh water has passed into the third or grilse state, and is then known as a grilse. Most of these fish run up from the salt water in the months of June, July, August, and September. The organs of reproduction are fully developed in both sexes; indeed, in the male this organ is found in the parr state to be perfectly capable of performing its functions.

There are still some persons who assert that the grilse never becomes a salmon, and that it is a distinct species. In 1863 a Committee of the Commissioners of the River Tweed Fisheries reported that, in their opinion, from the experience of the last twenty years, "a grilse never becomes a salmon at any stage whatsoever;" but three years later —viz., in 1866—the Commissioners published another report, and out of nineteen persons well acquainted with the river who were examined, only five stated that in their opinion a grilse never became a salmon. Two of the five gave their reasons. One was, that a kelt grilse marked in March or April was recaptured in the month of August a clean grilse; and Mr. G. Smith gave as his reason, "that he had never seen a marked grilse kelt return but as a grilse. The bones, scales, vertebræ, teeth, &c., of the grilse kelt are in general fully developed, while those of the spring salmon all denote youth, or those of young if not virgin fish."

Mr. Anderson, of Edinburgh, states also that a grilse never becomes a salmon. He says, in a letter addressed to the *Leeds Mercury*, "I have proven grilse a distinct species. The salmon and grilse have two distinguishing marks when only 1 lb. weight and up to 25 lbs.; of salmon, to 80 lbs. Salmon has a crescent-shaped tail and an oval scale, and the grilse a mackerel-shaped tail and a diamond scale. And another proof; we have salmon in January in our rivers 1 lb. and 2 lbs. weight, as well as salmon up to 80 lbs.; while the grilse never appear in our rivers till May, and these 1 lb. to 2 lbs., and the following year we find them 20 lbs. to 25 lbs. But the fishermen obtain a little more money in some quarters for salmon than grilse; all above 14 lbs. are called salmon."

We are afraid Mr. Anderson's so-called proofs will not satisfy many people. One fact is more conclusive than any amount of assertions, and when it has been over and over again proved by marking fish that most of the grilse return to the rivers the next year as salmon, diamond scales and mackerel-shaped tails go for very little.

Mr. Brown ("Stormontfeld Experiments") thinks that all the smolts of one year do not return the same year as grilse, the one-half returning next spring and summer as small salmon.

Day says: "It is very remarkable that grilse do not commence ascending until two or three months subsequent to the descent of smolts, whereas, had they been upwards of a year in the sea, it would appear strange why some few at least had not put in an appearance, this invariable absence from the nets almost seeming to point out the probability that they are not present."

Mr. Young states that he has often marked grilse, and that they have returned from the sea as salmon. A grilse kelt of 2 lbs. weight was marked on March 31, 1858, and was recaptured on August 2 of the same year as a salmon of 8 lbs.

Mr. Johnstone (evidence before Committee of the House of Commons), in describing the two fish, salmon and grilse, says: "The grilse is a much less fish in general, it is much smaller at the tail in proportion, and it has a much more swallow tail, much more forked; it is smaller at the head, sharper at the point of the nose, and generally the grilse is more bright in the scale than the salmon."

Day says: "Doubtless there is a difference in the appearance of the small salmon and a grilse of the same size, but such is probably due to the former, from some cause, not having got into condition, and so lost a season. That grilse frequently reascend rivers at irregular periods has been constantly observed, while they have also been entirely absent for a whole season, as in 1867."

It has also been observed that the grilse has more spots below the median line, and that the scales are smaller and more easily rubbed off, than the salmon.

Mr. Scrope ("Days and Nights of Salmon-Fishing") gives a table showing the increase in weight of grilse after returning from the sea as salmon, taken from an experiment made by a tacksman on the Duke of Sutherland's salmon-fishery on the river Shin. In the course of February and March 1841 he took a considerable number of grilse or gilses and marked them with a wire in various places sufficiently efficacious to be again recognised. Of these, ten were retaken in the course of the months of June and July following, by which time they had assumed the size and all the distinctive marks of the genuine salmon. It shows also how rapid the growth of the grilse is in his process of becoming a salmon :—

When Marked.	When Retaken.	Weight of Grilse.	Weight of Salmon.
		Lbs.	Lbs.
Feb. 18	June 23	4	9
,,	,, 25	4	11
,,	,, ,,	4	9
,,	,, ,,	4	10
,,	,, 27	4	13
,,	,, 28	4	10
March 4	July 1	4	12
,,	,, ,,	4	14
,,	,, 10	12	18
,,	,, 27	4	12

Willis Bund, in "Salmon Problems," gives another instance of the great and rapid increase of a salmon from his birth to his old age. "The late Mr. Ashworth," he says, "states at three days old a young salmon is nearly two grains in weight; at sixteen months old he has increased to 2 ounces, or four hundred and fifty times his first weight; at twenty months old, after the smolt has been a few months in the sea, it becomes a grilse of $8\frac{1}{2}$ lbs., having increased sixty-eight times in three or four months; at two and three-quarter years old it becomes a salmon of 12 lbs. to 15 lbs. weight; after which its increased rate of growth has not been ascertained, but by the time it becomes 30 lbs. in weight it has increased 115,200 times the weight it was at first."

Mr. Willis Bund remarks on this statement: "With the exception of the multiplication, one feels inclined to doubt the above statement, or at least to ask for the evidence on which it is based. Yet no one did more to bring about the present Salmon Acts, and they were mainly passed on such statements as these."

It is well known that a great number of salmon run up certain rivers very early in the year; indeed, they commence running in January and continue through February, March, and April. They are all in prime condition, generally large,[1] and it appears that a great proportion are

[1] In February and March a great many fine salmon from 15 to 43 lbs. are displayed on the fishmongers' slabs in London. We counted

male fish. There is no doubt that some of them run up as high as they can; for instance, most of the spring fish that enter the Ness pass at once up through Loch Ness into Loch Oich and up the Garry, where they afford capital sport in the early spring months.

Much has been written concerning these fish, and what is a spring fish is still in the realms of controversy. Why the salmon should select one river and not another close by is a mystery. On the west coast of Ross-shire there are two rivers, the Luing and the Elchaig, both running into Loch Luing in one stream, the Luing joining the Elchaig a few hundred yards before entering the loch. The Elchaig is an early river, plenty of fish running up in March and April, but not a fish goes up the Luing till much later. Dr. Day, "British and Irish Salmonidæ," has collected a number of facts concerning this subject, to which we would refer the reader.

It has been stated that salmon, after having entered the fresh water, return at times to the estuaries previous to the spawning period.

The late Mr. Robertson, head keeper on the Lochy, a most careful observer of the habits of salmon, informed us that he had often seen salmon returning to the salt water after a long spell of dry weather, when a freshet has set in, to have a wash, as he called it, and in a few hours ascend again; and when thus returning the large fish, particularly, are very fond of *showing themselves*, not by leaping, but by a kind of porpoise-like movement, a kind of roll. We once witnessed one of these second migrations as we stood waist-deep in a famous shallow, the fish surging on all sides, even running against our legs. There appeared to be hundreds of them, but not one would look at a fly.

It has been stated over and over again that salmon never feed in fresh water; that, having put on an abun-

---

eleven fish on one slab, three of them over 40 lbs., the rest from 12 to 18 lbs. These were all from Scotland. Seventeen salmon were taken in the Shannon by the rod in February 1889 ranging from 14 to 34 lbs.; twenty-five from the Tay ranging from 15 to 26 lbs.

THE SALMON.

dance of fat during their sojourn in the sea, they live upon that (like the bears in winter) during the whole time. Why do they, then, rush at and take a bundle of worms, or a gaudy fly, or a phantom minnow, or a young dace, or a boiled shrimp or prawn? Why do they take the boiled prawn, which they have never seen, and refuse the same in its natural colour? And why should they not live upon young fish, eels, &c., whilst in fresh water? They do not do so, say some, because nothing is ever found in their stomachs when taken. But it is a well-known fact that fish will vomit up the contents of their stomachs when in a

A FORTY-POUNDER "SHOWING HIMSELF."

fright; we have seen a sea-trout bring up half a dozen herring-fry. But not only that; many instances are recorded of food being found in the stomach of the salmon.

Parnell ("Fishes of the Firth of Forth") says: "I have repeatedly found the remains of worms and aquatic insects in the intestines of those salmon that were taken in rivers and lakes, but in those fish which were far advanced in roe both stomach and intestines were almost invariably empty."

Mr. Gosden, *Land and Water*, March 1886, states that he had examined the stomachs of 490 salmon, from the rivers Exe and Dart, and fifty peal (grilse?). In these he found eels, loach, minnows, gudgeon, sand-eels, shrimp, &c., an eel half-digested a foot long, and records a carp taken from a salmon caught in Hampshire waters. Another correspondent of the *Field*, February 20, 1886, a friend of Mr. G. M. Kelson, says: "I have seen salmon feeding in both rivers and lakes, and am simply astonished that any person should maintain they do not. A salmon caught at Kincardine had in its stomach seven sparlings, besides small shrimps; another, caught high up the Forth at Polnaise, contained a smolt and eighteen shrimps; one taken at Craiguith crieves, twenty-seven young eels; others have swallowed a trout fully half a pound and every imaginable insect, flies, beetles, worms, and spiders; so it is all nonsense to say that salmon when in fresh water live upon love. In 1844 two salmon caught in Loch Tay, in May, had in their stomachs one, and two young char quite entire, besides partially digested pieces of others."

Some of the fishermen on the Severn say that elvers are largely consumed by salmon, and on the Usk there is a saying, "A good year for prides (small lamperns), a good year for salmon."

It has often been stated that different rivers have their own variety of salmon. Robertson, head fisherman on the Lochy for many years, has told us that he could always distinguish a Lochy salmon from a Spean salmon. The fishermen of the Tay say that they can at once tell a Lyon salmon from a Tay salmon. The difference appears to be in the length or breadth of the fish.

Much has been written upon the leaping powers of the salmon. It has been stated by many writers that a salmon can spring perpendicularly from ten to fourteen feet.

Taylor ("Angling in All its Branches") says: "It is hardly credible by those who have not witnessed it that these fish will leap full twelve feet perpendicularly—nay, allowing for the curvature, they must sometimes leap sixteen to eighteen feet."

Cholmondeley Pennell ("Angler Naturalist") says the limit certainly does not exceed twelve or fourteen feet, but much depends on the depth of the water from which the salmon makes its spring.

Stoddart ("Angler's Companion") thinks they cannot surmount falls exceeding twelve feet.

Scrope ("Days and Nights of Salmon-Fishing") says that, from personal observation, he never saw a salmon spring out of water above five feet perpendicularly; and Yarrell says that eight to ten feet is the limit. Then, again, Twiss ("Travels in Ireland") states that he has seen salmon dart themselves nearly fourteen feet perpendicularly.

Most probably in all these statements none of the obstacles or falls have been accurately measured, and the supposed height has been all guesswork. But in the following, taken from *Nature*, we have much better evidence, the fall having been carefully measured:—

"SALMON LEAPS.—Professor A. Landmark, Chief Director of the Norwegian Fisheries, has published some interesting particulars of his studies of the capability of salmon to jump waterfalls. He is of opinion that the jump depends as much on the height of the fall as on the currents below it. If there be a deep pool right under the fall, where the water is comparatively quiet, a salmon may jump sixteen feet perpendicularly, but such jumps are rare, and he can only state with certainty that it has taken place at the Hellefos, in the Dråms river at Haugsend, where two great masts have been placed across the river for the study of the habits of the salmon, so that exact measurements may be effected. The height of the water in the river of course varies, but it is, as a rule, when the salmon are running up-stream, sixteen feet below these masts. The distance between the two is three and a half feet, and the Professor states that he has seen salmon jump from the river below across both masts. Professor Landmark further states that when a salmon jumps a fall nearly perpendicular in shape, it is sometimes able to remain in the fall, even if the jump is a foot or two short of the actual height. This, he maintains, has been proved by an overwhelming quantity of evidence. The fish may then be seen to stand for a minute or two a foot or so below the edge of the fall in the same spot, in a trembling motion, when, with a smart twitch of the tail, the rest of the fall is cleared. But only fish who strike the fall straight with the snout are able to remain in the falling mass of water; if it is struck obliquely, the fish is carried back into the stream below. This, Professor Landmark believes to be the explanation of salmon passing falls with a clear descent of sixteen feet; the Professor thinks that this is the extreme jump of which salmon are capable, and points out that, of course, all have not the power of performing this feat."

The power of making the spring is centred in the strong muscles which regulate the tail, and this is exerted to the utmost at the moment of the attempt. As late as the year 1822 the old myth described by Drayton, that salmon made the leap by putting their tails into their mouths and using it like a bow, was still believed by some authors.

Williamson's "The Complete Angler's Vade-Mecum," published in that year, states that he has frequently seen them ascend in this manner about ten or twelve feet, and has read of them leaping much higher.

If salmon are carefully observed when leaping at a fall, it will be seen that many of those which fail to get over, as they fall back have their bodies curved. This may account for the above strange notion. The fish which get over a high fall either have extra power and have taken the leap perfectly straight, or have been assisted by an unseen ledge in the falling water, by which they are enabled to use a fresh muscular effort. We have often seen salmon shoot up in the middle of a fall after taking the leap of some eight feet or so, and we have also noticed that salmon will jump for a particular rock protruding from the side of a fall with only the smallest quantity of water running over it; and the moment the fish touched it, with another spring the remainder of the fall was surmounted.

Salmon grow to a great size. The largest authentically recorded to have been taken in this country weighed 83 lbs., and was exhibited at Mr. Grove's, fishmonger, in Bond Street, in 1821.

THE SALMON.

The largest salmon taken in the United Kingdom with *the fly* is stated to have weighed 57 lbs.

The salmon has been subject to certain laws from a very early date. In the days of Howel the Good, who died in 948, there were three common hunts (see "Hunting Laws of Cambria"), a stag, a swarm of bees, and a salmon. Salmon are called a common hunt because when taken in a net, or with a fish-spear, or in any other manner, if any person whatever comes up before they are divided, he is entitled to an equal share of them with the person who caught them, if it be in common water.

In Holinshed's "Chronicles, England," vol. i., Book iii. chap. 3, in an account of the fish in our rivers, it says: "Besides the salmon, therefore, which are not to be taken from the middest of September to the middest of November, and are verie plentiful in our greatest rivers, and their young store are not to be touched from mid April unto midsummer, we have trout, barbel, grilse, powt, chewin, pike, goodgen, smelt, perch, &c., whose preservation is provided by verie sharpe lawes." It says, in giving the names of the fish, also, that a salmon is the first year a gravellin, and commonly as big as a herring, the second a salmon peal, the third a pug, and the fourth a salmon. The spawning-time is mentioned as much earlier than at present :—" The salmon in harvest-time cometh up into the small rivers where the water is most shallow, and there the male and female, rubbing their wombe one against another, they shed their spawne; if they touch anie of their full fellows during the time of this leannesse, the same side which they touch will likewise become leane, whereby it cometh to pass, that a salmon is oft seene to be fat on one side of the chine and leane on the other."

And as to their leaping at falls : " Such as assay often to leap and cannot get over, do brooze themselves and become measelled—and the people set up a cauldron of hot water upon the shallows in hopes to catch the fattest, as such by reason of their weight do oftenest leape short."

"The Donne and the Dea," he says, "give the greatest salmon that are to be had in Scotland; and of the River

Wie or Guie, it is found by common experience that the salmon of this river is in season when the like fish to be found in all other rivers is abandoned and out of use, whereof we of the east parts doo not a little marveil."

Of the Thames he says: "What should I speake of the fat and sweet salmons dailie taken in the streams of this noble river, and that in such plentie, after the time of the smelt be past, as no river in Europe is able to exceed it. What store, also, of barbel, trouts, chevins, pearches, breames, roches, daces, gudgeons, flounders, shrimps, &c., as are commonly to be had therein!" And then he says: "From the insatiable avarice of fishermen, how this famous river is defrauded, yet complaineth of no want, but the more it loseth at one time, the more it yieldeth at another; only in carps it seemeth to be scant, but it is not long since that kind of fish is brought over to Englande." And he concludes thus: "Oh that this river might be spared but even for one year from nets."

THE SEA-TROUT.

The SEA-TROUT (*Salmo trutta*). There is a difference of opinion amongst ichthyologists and others as to whether there is only one, or whether there are more than one, species of sea-trout indigenous to the British waters.

THE SEA-TROUT.

Are we to consider *Salmo trutta* and *Salmo cambricus* as two distinct species, or are we to take Dr. Day's view, that these two fish are one and the same, dividing them into a northern and a southern race?

Are we, as some suppose, to consider the phinnock or herling a distinct species, or only the grilse state of the sea-trout?

Is the whiting or whitling of Cumberland the same as the phinnock or herling?

Is the white trout of Ireland a distinct race, or the same species with the *Salmo trutta*?

After a very careful study of the mass of information collected on this subject, it appears to us that the external distinctions which have been relied on are dependent on localities and other causes which are liable to change; that the internal differences are also too variable and inconstant to form specific characters; and that the so-called species, as Day says, pass one into another by insensible gradations, without showing any line of demarcation; that, if anything, they are merely local varieties. Indeed, some zoologists hold to the opinion that the sea and fresh-water species are merely local races of one species; that the anadromous and fresh-water forms simply result from local circumstances consequent on immediate surroundings, but that both are descended from one ancestral form.

Major Treherne, in his evidence before the Commissioners appointed to inquire into the English and Welsh salmon-fisheries, stated that in his belief the sewin of Wales, the sea-trout of Scotland, and the white trout of Ireland are the same fish; there is also reason to suppose that the salmon trout and peal of Devonshire and Cornwall are the same as *Salmo trutta*, and that the phinnock, herling, and whitling are only local names for the grilse state of the sea-trout, *Salmo trutta*. Much has been written on this subject, many believing that the phinnock and herling are distinct species. Day collected an immense amount of information, all tending to prove that Major Treherne is right in what he states.

The parr of the sea-trout—orange-fins, as they are called—is marked almost similarly to the parr of the salmon, but the dorsal fin has generally a more distinct white upper edge anteriorly and a blackish basal band. Its adipose fin is tipped with orange. In the grilse state the tail is dark, growing lighter as the fish gets older; the spots extend below the lateral line.

In the mature sea-trout the cross spots extend as low as the pectoral fins, and in the breeding season, after sea-trout have been in fresh water for some time, in some examples a few red spots, similar to those found in *S. fario*, appear along the lateral line, particularly in the male, and the silvery abdomen becomes yellowish.

The body of the sea-trout is not so elegant as that of the salmon. The fins in the larger specimens are somewhat shorter than in the smaller. The ventral fin is on a line beneath the middle or last third of the dorsal; dorsal somewhat small. The caudal (tail) is very variable in form, according to age, &c., sometimes square, sometimes notched, and sometimes convex. The teeth in the young parr are in a double line, but as they get older they become a single line along the vomer, with a transverse row across the posterior edge of the head of that bone.

Fin-rays: dorsal, 12–14; pectoral, 13–14; ventral, 9; anal, 11–13; caudal, 19–21.

This fish, either under the name of sea-trout, sewin, or white trout, gives almost as much exciting sport, owing to its gameness and courage, as *Salmo salar* himself. It has been truly said that there is no fish that swims which rises so boldly at the fly, and when hooked shows such fighting powers and such courage. Taking size for size, the sea-trout shows more pluck than the salmon, and there is one peculiarity in favour of the former, he seldom sulks when hooked. He is, as Francis Francis says, "like a champion of light weights, here, there, and everywhere; now up, now down, now in the water, now out." He fights to the last, even when in the net. A fresh-run sea-trout of from 2 lbs. to 4 lbs. is worth fishing for.

The general size of a sea-trout is from 1 lb. to 7 lbs. We have seen some from 10 lbs. to 16 lbs. The colour of the flesh is of a peculiar pinky-red, very different to the salmon, and quite different to that of the bull-trout, which is of a yellowish hue with a shade of pink; and whilst the

THE SEA-TROUT.

one is firm and savoury to the taste, the other is "leather-like and insipid."

The bull-trout, *Salmo eriox*, as it is called, is by some considered as a distinct species. Others assert that the sea-trout and bull-trout are identical. Buckland says, whether identical or not, he could pick out a bull-trout from a thousand of the other kinds of *Salmonidæ*. Between the two fish all the Scotch fishermen consider there is a marked difference. Buckland says: "I recollect hearing of some Scotch fishers who had caught a curious-looking fish. 'It's just a kelt,' said one. 'It's a bull-trout,' says another. 'It's a sea-trout,' says a third. 'What's the odds,' says the tacksman, 'what you call him? He'll just be a saumon, worth 1s. 6d. per pound in the London market; shove him into the ice-box.'"

Day places the bull-trout as the northern form of sea-trout, and Günther considers *Salmo eriox* and *cambricus* the same.

Mr. Kerr, writing to Dr. Day, has made some very interesting observations in relation to sewin and bull-trout, and he sends a specimen of the fish he caught in the river at Tan-y-Bwlch. He says: "It is a fair specimen of what they call here *sewin*, which I take to be the Scotch sea-trout. I have caught lots of fish precisely like this one in the rivers flowing into St. Andrews Bay, and from there down to the English Border, and we always imagined (though, it may be, wrongly) that they were all sea-trout. There is a fish that I have caught in the Tweed and Teviot also very similar to this one, and called there the *bull-trout;* here they give that name to yellow trout that have gone down to the sea-water, but the fish I refer to, which I have so frequently caught, but only in Tweed and Teviot, is a true migratory species, and one which has often puzzled me. I have caught a good many sewin in this river this summer, and it has struck me that most of them were much more like, both in appearance and when on the table, these so-called *bull-trout* of the Border than the ordinary type of Scotch sea-trout that I have caught in the Highlands and elsewhere."

It appears to us that the fish is distinct from *Salmo trutta*, but as the term is constantly applied to the various forms of sea-trout, it would be of importance to ascertain if the so-called bull-trout of other localities is precisely the same as the bull-trout of the Tweed.

The true bull-trout should be known under one specific name, as *S. eriox*, and by that only; and the so-called bull-trout of other rivers should be either identified with this or placed under *S. trutta*.

THE RIVER-TROUT.

The RIVER-TROUT (*Salmo fario*). Eight or nine different species or varieties of British fresh-water trout are described by ichthyologists.

Dr. Günther, "Catalogue of British Fishes," divides *Salmo fario* into two races, a northern and a southern.

The first he calls *S. fario gaimardi*, and places its habitat in North Britain and Ireland, giving examples from the Tweed, the Esk, the Clyde, and from various rivers and lakes in Ireland.

The second he calls *S. fario ausonii*, placing its habitat south of the Tweed; giving examples from Cumberland, Westmoreland, Wales, Shropshire, Hampshire, Buckinghamshire, Surrey, and the Thames.

Dr. Günther's species are *S. fario gaimardi*, *S. fario ausonii*, *S. orcadensis*, *S. ferox*, *S. stomachicus* (the Gillaroo), *S. nigripinnis*, *S. levenensis*, and two closely allied to the non-migratory forms, *S. brachypoma*, which by some ichthyologists is considered to be identical with the phinnock or grilse state of the sea-trout, and *S. gallivensis*, identical with the estuary trout of Knox.

Dr. Day gives one species and eight varieties, viz., *Salmo fario*, with Loch Leven trout, Grasspuil trout, Estuary trout, Orkney trout, Cornish trout, Great Lake trout, Gillaroo trout, and Swaledale trout, as varieties.

Yarrell gives three species, *Salmo fario*, *S. ferox*, and *S. levenensis*.

The river-trout varies in colour and shape according to the river which it inhabits. This variation arises in a great measure from local circumstances. Trees or bushes overhanging the stream, gravel or mud in its bed, great or less amount of light, depth of water, distribution and quality of food, all go to alter the condition and aspect of the fish.

The trout in one part of the river may be silvery and marked with many dark cross spots, and with very red flesh; whilst in another part they are yellow-sided, red spotted, and with white flesh. Change the fish from one to the other locality, and they in a very short time take on the local form and colour.

In the year 1867 two separate lochs in Scotland were stocked with trout from Loch Morar, both lochs being about 1500 feet above the level of the sea. In the one, possessing a sandy and weedy bottom, we caught, in 1882, a great many trout up to a pound weight. They were long lanky fish, their sides and under-surface having a golden tinge. The spots were numerous and red, and the flesh white.

In the other loch, not two miles away—smaller, not very deep, with a very rocky bottom—the fish do not exceed $\frac{3}{4}$ lb. to 1 lb. in weight, are very dark in colour, with yellow sides, covered with purple and red spots, and the flesh pink. So altered were these fish from those caught in Loch Morar, owing to changed conditions of life, that they might easily be considered as distinct species.

" In the streams and cuts through the flat bogs below the high land, there are large, blackish fish, locally known as bog-trout. In the spring of the year, when the bog-holes are tapped, these are half-smothered, and float on the surface of the water, and are caught by the country-people, the streams or cuts being often depopulated; yet before the next spring the cuts again have their denizens, so that the fish which come down from the hills not only increase in size, but change their colour to suit circumstances. This change of colour is very remarkable in some of the Connemara streams. In the boggy portions 'bog-trout' are found, and in the others, above and below, 'silver-trout.'"

Many other instances might be mentioned. Yarrell says: "When we consider geologically the various strata traversed by rivers in their course, the effect these variations of soil must produce on the water, and the influence which the constant operation of the water is likely to produce upon the fish that inhabit it; when we reflect, also, on the great variety and quality of the food afforded by different rivers, depending also on soil and situation, and the additional effect which these combined causes, in their various degrees, are likely to produce, we shall not be much surprised at the variations both in size and colour which are found to occur."

The teeth in *S. fario* are in a double row on the vomer, numerous, strong, and curved inwards.

Fin-rays: dorsal, 12–15; pectoral, 13–15; ventral, 9; anal, 10–12; caudal, 19.

The river-trout is the prince of fresh-water fishes. Yarrell describes it as "vigilant, cautious, and active;" Day, as "bold, voracious, cunning, and shy; possessing keen sight, and appears to be suspicious of anything novel." We would give the trout a far better character, and describe it as bold and determined, wary and watchful, with its perceptive faculties much more developed than any of its race.

All living creatures which possess a brain, after coming in contact with the human race generally become very shy and wary. This has been particularly noticed with mammals and birds; then why should it not be so with fish?

The river-trout is the only species of our British *Salmonidæ* which when on the feed lies very near the surface of the water, and so becomes much more liable to be disturbed by passing objects, and any unwonted circumstance throwing a shadow or causing reverberation will what is called send a trout down. In these days, with so many rod-fishers, especially those who practise the art of fly-fishing, most of the feeding fish are at one time or another thrown over, either by the fly or by an artificial minnow being presented to its gaze.

In earlier days these were taken as realities by the fish and seized; as years passed on, less and less notice was taken of these lures, especially in much-fished rivers; the perceptive faculties of the trout became more acute, and were transmitted by hereditary descent to their progeny. The power of discriminating the real from the unreal is becoming more and more apparent, and the power of capture more and more difficult.

Our prophecy made in 1884 (see Preface to "Recollections of Fly-Fishing," &c.) is being fulfilled. Some think that more transparent and finer gut-casts must be used; others, that a more true imitation of the exact fly must be made. As a rule, artificial flies are not true imitations—too many legs, too many whisks or setæ, too thick bodies, which are not sufficiently transparent, &c., &c. All these must be modified, and an exact likeness of the natural fly must be placed before the fish, if the fly-fisher wishes to keep the upper hand.

We do not lay so much stress on fish having been hooked and got away, or of remembering the circumstance for any length of time; as far as one can judge, the faculty of the memory is not very strongly developed; so many instances occur every year of fish being caught, some even a few minutes after a struggle for life, and many after a few hours, with the instrument of capture still sticking somewhere in their jaws. But we believe in the increase of other perceptive faculties, particularly in that of vision, and, through that, of knowledge of what is right and wrong, and of what may be on the bank—whether an enemy in the shape of man, or a harmless intruder in the shape of a cow or horse.

The powers of reproduction commence about the second year. The spawning season commences in November and lasts till January, dependent on climate and locality. Some rivers are early; some trout will spawn in October, and some as late as February; but, as a rule, it is at its height from the middle of November to the middle of January. Spawned trout take some time to get into condition. It would be far better for the fish if the close-time continued till the 14th of April at the earliest. There are some fish always in order, but these, we suspect, would be barren fish, *i.e.*, wanting in the organs necessary for the reproduction of their race. This abnormal condition is found in every class of animal life.

All the salmon family have fine qualities and afford grand sport. They are all, when in season, of elegant shape and handsome appearance. We admire the anadromous forms for their blue steel-coloured backs and silver sides; but of all the non-migratory inhabitants of the fresh-waters, there are none to be compared to the

" Trouts bedropped wi' crimson hail."

The Loch Leven trout, *Salmo levenensis*, has had particular attention drawn to it from its successful cultivation at Howietoun, at Mr. Andrews' fish-culture establishment at Guildford, and elsewhere, for the purpose of introducing it into the lakes and rivers, not only of this country, but in all parts of the world where the *Salmonidæ* can live. In the *Fishing Gazette* of December 1886 we entered fully into the history of this fish, and again in a paper written for the Glasgow Trout Preservation Association. It appears, from good authority, that its introduction into some rivers has greatly improved the quality of the native trout. If this should continue, it may help to solve the problem as to whether the *S. levenensis* is a land-locked sea-trout, or whether *S. trutta* and *S. fario* are of the same species.

As regards the great lake-trout, *S. ferox*, most authors and fishermen who have studied the subject, we believe, have come to the conclusion that *ferox* is nothing but an overgrown ugly example of *S. fario*.

Much has been written on the subject of restocking trout-rivers. Of late years many of our trout-streams have been suffering from a maximum of fish and a minimum of food, and, as a natural consequence, deterioration in the growth and in the quality of the trout.

There are some very marked causes for this state :—

The result of overstocking, from the desire of many owners of fisheries to have plenty of trout.

The preservation of old fish, male and female.

Too many trout in a stream, with only a certain quantity of food-supply, naturally alters their condition and size. They become long and flabby, lose their energy and boldness, and are miserable objects to look at.

The preservation of the old fish has the same tendency; fish after a certain age—earlier, we believe, than is generally supposed—lose their power of reproducing healthy stock; their progeny is weak and nothing like so strong and healthy as those of parents in the prime of life. It is, then, of importance to breed as much as possible from fish in full vigour of life, which, we should think, would be about the fourth or fifth year.

Some think that female trout should not be killed towards the close of the season. This is certainly a good rule if it is confined to young females or those in their prime. But old females, as well as old males, are better out of the water. Trout are in a great degree polygamous, and, like all polygamous vertebrates, their power of reproduction is, to a certain extent, limited. If it is necessary to limit the number of females to each male in all polygamous mammalia and birds, if healthy and strong offspring are to be produced, why should not the same rule apply to fish?

A day with the trout on the spawning-beds is replete with interest. We passed such a day on the 14th of January of this year. Although not very favourable for observation, there being no sun, yet the water was so shallow and so clear that all the movements could be plainly seen. A great many rhedds had been already made, and a large number of fish had already spawned; yet in a space certainly not more than forty yards in length and six yards in breadth, there were at least 300 brace of trout hard at work. The process appeared to be as follows :—A female, or more correctly speaking a hen-fish, would take up her place in one of the depressions formed in the gravel, and close behind her were generally two male or cock fish (sometimes as many as four or five). These remain perfectly quiet until the hen-fish suddenly turns on her side and sheds her eggs in the hollow of the gravel. As soon as this is done the cock-fish commence to fight, and the victor, having driven his rival away, places himself over the place vacated by the female, deposits its milt, and then the female having returned, they both work with their tails and lower part of the body to cover the deposited eggs with a coating of gravel and stones, some of these stones being of considerable size. This process is going on all over the shallow occupied by the trout. The fights between the cock-fish are constant and continual, and the turmoil that goes on would, one would think, utterly disturb the females; but evidently they like it, and the more lovers they have, the better they appear to be pleased. We noticed many of the male fish gashed and torn in these battles; one or two appeared to be severely wounded, and would probably not recover. At this time it is easy to calculate the number and the size of the fish in the river, for every shallow in the whole length of the water, some three to four miles, was almost as fully populated. But the large males are not always the accepted lovers, being ousted by younger and more vigorous fish; they rush about from one rhedd to another, endeavouring as much as possible to annoy and harass their more successful competitors. The female fish are generally of a lighter colour than the male, and are more difficult to see; and Buckland states, on his own personal observation, that the females by some means or other manage to elude the nets, and seek safety by getting into holes under weeds, and among the roots of pollard-willows under the banks. He gives several instances of the craftiness of the females in eluding, and the blind rush of the males in rushing into, the nets

THE RIVER-TROUT.

(see "Familiar History of British Fishes," p. 292 *et seq.*). It certainly is the case that in catching trout for spawning purposes many more male fish are caught than females. Buckland's suggestion may account for this.

As to the other so-called species of fresh-water trout, great difference of opinion still and probably always will exist. The great lake-trout, *Salmo ferox*, for instance, is considered by the late Dr. Day to be merely an old and large *S. fario*. Yarrell makes it a distinct species, as having thirteen dorsal rays instead of fourteen, with a different form of scale. Günther makes a difference of vertebræ—56–7 in *S. ferox*, 59–60 in *S. fario*. But Day asserts that there are undoubted specimens of *S. fario* with 13–15 rays on dorsal fin, and with 56–60 vertebræ.

The *Salmo stomachicus* (the Gillaroo), has been made into a species, from the membranes of the stomach being so much thicker than in any other trout; the young fish do not show this peculiarity, and appear to feed exactly on the same shells, *Limnæus*, *Ancyclus*, &c., which other trout eat, without forming gizzard stomachs.

Again, the variety or species which under the name of Loch Leven Trout has had particular attention drawn to it from its successful cultivation at Howietoun, Guildford, and other fish-culture establishments, is perhaps more puzzling. It has been for very many years a subject for argument and difference as to whether this fish should be considered a species distinct from *S. fario*, and, if so, whether it might not be a descendant of a marine form (*S. trutta*), which, having ascended into Loch Leven from the sea, was suddenly prevented returning; and if this should be the case, whether its introduction (provided *S. trutta* and *S. fario* are distinct races) into rivers and its interbreeding with *S. fario* might produce hybrids, and thus tend to sterility. For the present, it appears, from the observation of different people, that this is not to be dreaded. Mr. Spalding states that the introduction of Loch Leven trout into the river Darent in Kent, has greatly improved the quality of the trout, their flesh, from being white and flabby, becoming pink and firm.

Sir Humphry Davy, Preface to "Salmonia," 1828, says, in writing of Loch Leven trout: " If trout from a lake or another river of a different variety were introduced into this river (the Teme), they would not at once change their characters, but the change would take place gradually. Thus I have known trout from a lake in Scotland, remarkable for their deep-red colour, introduced into another lake where the trout had only white flesh, and they retained the peculiar redness of their flesh for many years. At first they all associated together in spawning in the brook which fed the lake, but those newly introduced were easily known by their darker backs and brighter sides. By degrees, however, from the influence of food and other causes, they became changed; the young trout of the introduced variety had flesh less red than their parents, and in about twenty years the variety was entirely lost, and all the fish were in their original white state." No form of trout alters so much from its original form as *S. levenensis* when placed in other waters. Examples have repeatedly come under our notice where in the course of three years these fish have so entirely changed that it would be difficult to recognise them as the beautiful fish of the Fifeshire loch. Should, however, repeated experiments and proportionate lapse of time prove that the introduction of the Loch Leven trout into our rivers does permanently improve the race (believing, as we do, that this trout is the descendant of the anadromous form), it will go far to confirm the assertion of Widegreen, Collett, and others, " that there is only one species of trout," and since advocated by Dr. Day, who says "that he is unable to accept the numerous species which have been described, believing those ichthyologists more correct who have considered them modifications of only one, which, as *Salmo trutta*, includes both the anadromous and non-migratory fresh-water forms."

THE GRAYLING.

The GRAYLING or GREYLING (*Salmo thymallus*) is placed by Günther after the genus *coregonus*, and *thymallus* appears to be more closely allied to the gwyniad and vendace than to *Salmo fario*.

The head and body of the grayling is long and elegant; the sides have longitudinal bands; the large dorsal fin, much longer than high, has numerous rays and a number of spots of a beautiful purple; mouth small, with the upper lip slightly overhanging; teeth very small on the jaws, near the head of the vomer and on the palatines, none on the tongue; the tail is forked. This fish, when in proper season, is one of the handsomest as well as one of the most graceful of the family. The small, dark olive head,

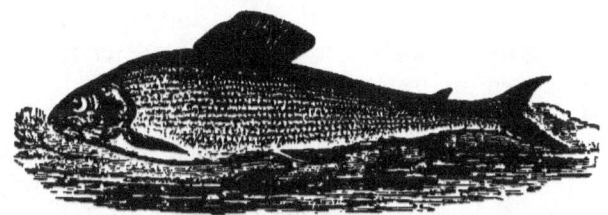

THE GRAYLING.

with the peculiar pear-shaped pupil of the eye, surrounded by its parti-coloured iris of green and purple, enclosed in a ring of golden yellow, the gill-covers iridescent with numerous tints as you expose them to the angles of light, the great purple dorsal fin, the golden-yellow pectoral and ventral fins, the forked tail or caudal fin, with the two long golden streaks extending on each side of the pectoral fins, the dark olive, blackish-green of the back, the sides glistening with greyish-pink, silver lines, tipped with gold, the white abdomen and throat, make up the very perfection of a beautiful fish. The great dorsal fin is the peculiar feature when looking at a grayling; to see this properly it should be stretched out and held up to the light. The lovely purple-red and dark bars with the purple streaks along the rays have a singularly beautiful appearance, likened by Mr. Pritt to the wing of a butterfly.

Fin-rays: dorsal, 20–24; pectoral, 15–16; ventral, 10–11; anal, 11–14; caudal, 21.

The specific name *thymallus* is said to have been given by Ausonius from its peculiar thyme-like smell when first captured. This has been doubted by many. Some say it has no smell at all; others that the scent is more like cucumber; whilst Mr. Lloyd is inclined to attribute it to their eating the *Gyrinus natator*, the water-bug, a little water-insect which has the most unpleasant odour, very different to that of the grayling.

This peculiar thymo-cucumber scent of this fish is most perceptible in fish from $\frac{3}{4}$ lb. to 1 lb. in weight.

In former days the grayling was known under the name of Umber, derived, according to Cotton, from its blackish-brown spots, but according to other authorities from the Latin word *umbra*, a shadow, in allusion to its rapid rise and its sudden disappearance.

The term grayling is said to be a corruption of "grey lines." Might it not be from "gray-ling," the latter being a corruption of "long," or "slender," in the same way as the *Neotrus vulgaris* is called the Ling or Long?

Grayling have been supposed by some to have been imported from the Continent, but there is very slender foundation for this, except that the fish are found in rivers on which religious houses were placed. In those early days, when fish was an important article of the diet of the monks and those professing the Catholic religion, it is much more probable that the monasteries and religious houses were built on the banks of rivers where the fish, especially trout, grayling, and pike, were most prevalent; and as the grayling comes into season when the trout is unfit for the table, it would be of considerable importance to all religious orders to have this fish near them. In those days there would be considerable difficulty, to say the least of it, of importing the fish from the Continent, and artificial spawning was not in vogue at that time.

THE GRAYLING.

We can see no reason why the grayling should not be an indigenous fish.

The proper season for grayling is from the beginning of September to the end of January. They are then in their prime, and are bold and daring, rising again and again at the fly.

Some consider the grayling interferes with the trout, not only by devouring the spawn, but by taking all the flies on the top of the water and all the larvæ at the bottom. They liken this beautiful and sporting fish to a pig, and, "like a pig in his habits, he is particularly fond of rooting up and feeding on the bed of the river, picking up the abundant harvest they find there in the shape of aquatic insects in their larval states."

From the very tender mouth of this fish, it may be doubted if it can rout up the rhedds made by the trout. Like all fish, even *S. fario* itself, they will take the ova that float down the stream, most of which are probably unprolific.

Why attack the grayling for eating the larvæ at the bottom? What are the trout doing when they are what is called tailing? and when you compare the mouths of the trout and grayling, which will do the most damage?

Trout, when feeding on the fly, lie just below the surface, so that they can take their prey with very little effort, and any one who will have the patience to watch a trout feeding, and resists the desire to throw over him for a time, will see what a great number of flies pass him without the slightest notice being taken of them. Grayling lie at the bottom, and have to rise rapidly at the fly, and at once return to their haunts. Which of the two has the best chance? We feel quite sure that trout and grayling will live quite happily together. The fault lies in overstocking a river, when too many fish are preserved for the quantity of food supplied. In the present day everything must give way to numbers.

All admire the purple-spotted dorsal fin, which, when fresh from the water, is more like a beautiful piece of fine tortoise-shell. To what purpose, in the economy of this fish, is this huge fin, so different to all the other members of the salmon family, applied? We have had the opportunity of being able to watch large grayling when completely at rest in a deep pool on the river Test, where they were living in the greatest amity with a number of large trout. As long as the grayling were quite stationary, the great dorsal fin was lying folded over on one side of the back, generally, if not always, on the left. When the fish desired to rise to the surface of the water the fin was very slightly moved, but when about to descend again it was raised to its full extent, and it appeared as if its purpose was to enable the fish to descend with great rapidity. The fin probably on being raised acts on the muscular fibres connected with the air-bladder, rapidly compressing it, and thus increasing the quick descent so peculiar in this fish.

All who have watched a grayling after being hooked will observe with what tenacity he endeavours to get to the bottom, and how rigid and upright the dorsal fin continues during the fight.

THE CHAR.

The sub-generic group, *Salvelini*, contains the CHAR or CHARR (*Salmo alpinus*). This fish inhabits the deep lakes of Westmoreland, Cumberland, and Lancashire, some of the lakes in Wales, and many of the lakes in Scotland.

The British char has been divided by some writers into six species; Day considers there is only one, with five varieties. The principal feature in this sub-genus is, that there are no teeth down the body of the vomer, but only on the head of that bone. The different species are distinguished by colour, development of the maxilla, size of the teeth, difference in the fin-rays of the dorsal fin, cæcal appendages, the size of the scales, and number of the vertebræ.

THE CHAR.

In Windermere there are two kinds of char known to the fishermen—the silver and the gilt char, the distinction resting entirely on the difference of colour and quality. Dr. John Davy ("Angler and His Friend") says: "Were the naturalist to attend to colour and spots or markings, as many species of char might be established as there are localities in which it has been found, inasmuch as the char of each lake has, as regards colouring and spots, something peculiar."

Dr. Günther's division of the British char is as follows:—*Salmo alpinus*, Scotland; *S. Killinensis*, Loch Killin, Scotland; *S. Willoughbii*, Windermere Lake,

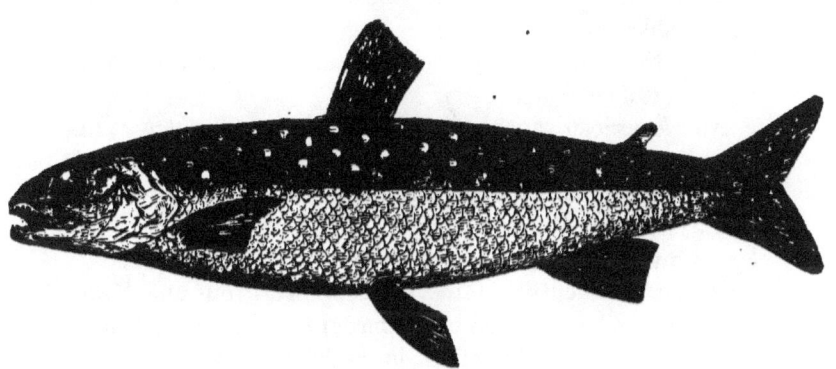

THE CHAR.

Westmoreland, and Loch Bruach, Scotland; *S. Perisii* (the Torgoch), lakes of North Wales; *S. Grayii*, Loch Melvin, Ireland; *S. Colii*, Lough Eske and Lough Dan, Ireland.

Day gives *Salmo alpinus* as the British char, with five varieties, which are the same as Günther's species.

None of the British char appear to be migratory.

Char are gregarious, and for the most part deep-swimming, occasionally coming to the surface in bright, hot weather. In the autumn months they frequent the streams for the purpose of spawning, and it is then that they are often taken with the fly, when fishing for sea-trout in some of the Scottish lochs. In the Lake District they are caught in considerable numbers by deep fishing with a worm or minnow.

Agazziz states that our *Salmo alpinus* is identical with the *Ombre chevalier* of the Lake of Geneva.

The char (*S. alpinus*) is a very handsome fish as to shape and colour. The colour varies, but as a general rule the lower parts of the abdomen and sides become—prior to spawning—of a deep scarlet; hence its Welsh name, *Torgoch*. The spots on the upper part of the body are orange-red, with some dark-coloured; the top of the head and upper part of the back umber-brown; pectoral, ventral, and anal fins deep reddish-orange; dorsal and caudal fins dark-brown, with a purplish tinge; gill-covers yellowish-olive; sides orange.

Fin-rays: dorsal, 13; pectoral, 12; ventral, 9; anal, 11–12; caudal, 19–21.

An American species of char, *Salmo fontinalis*, has been introduced into this country with variable results. It appears from recent observations to be, like some of the northern species, according to Pallas, a migratory fish. Dr. Brown Goode ("Game-Fishes of the United States") says that the identity of the Canadian sea-trout and *Salmo fontinalis* has been settled beyond a doubt. Mr. Armistead, of the Solway Fishery, Dumfries, in the *Fishing Gazette*, January 26, 1889, states that his fry of the *S. fontinalis* have been caught in the Solway, weighing several pounds; and in a letter to the same of January 12 says that this fish "has been put into several lakes and streams in Galloway, but has in all cases bolted and 'run away to the sea,' some being unable to pass the salmon stake-nets, and these have varied from 4 lbs. to 12 lbs." Mr. Armistead considers it an anadromous fish. Should this be the case, it will make this fish a very desirable addition to our rivers, and it would be most interesting to discover under what conditions those which have "run away to the sea" have been seen or captured on their return to fresh water.

THE SMELT.

The SMELT (*Osmerus eperlanus*), Sparling in Scotland, is not often an angler's capture, although Dr. Day states they may be caught with a paternoster line, with No. 8 or 9 hooks baited with shrimps or gentles, red worms, &c. Salter, in his "Angler's Guide," says he has caught very fine smelts by angling in Portsmouth Harbour, but Yarrell has no doubt that these fish were the sand-smelt or atherine, one of the *Mugilidæ* (mullets). The smelts are usually caught in nets in the estuaries of rivers, Medway smelts being famous. They are very plentiful along the east coast. They enter the rivers in August for spawning purposes, and return to the sea in May. The fish has a very peculiar cucumber odour when fresh taken. Smelts have been bred in fresh water having no communication with the sea, and have thriven well, propagating abundantly. It spawns in March. The smelt is found in the Tay, the Forth, the Ure, the Humber, the Thames, and the Medway, on the east coast; on the west, in the Solway Firth, and as far south as the parallel line formed by the Mersey, the Dee, the Conway, and Dublin Bay. Yarrell states that it is not found between Dover and the Land's End.

In Salter's day ("Angler's Guide," 1814) smelts were comparatively common in the Thames, although the numbers had greatly fallen off. He says: "They arrive twice a year in the Thames—March and July—but do not go above London Bridge in their last visit. The most favourite resort for taking these fish in nets was between London Bridge and Lambeth. The river formerly swarmed with this delicious fish. In July we used to begin angling for smelts in the various wet-docks below bridge, but the spot covered by the floating timber at Limehouse Hole was a favourite resort of the smelt-fishers. A hundred dozen have frequently been taken in a day. I have made much inquiry on the subject of smelts not visiting as usual the River Thames, but without any satisfactory results."

Captain T. Williamson ("The Complete Angler's Vade-

Mecum," 1822) says that smelts never stray far from salt water, or, at least, where it is a little brackish; that they come in with the flood, especially during spring-tides, and return with the ebb. They are fond of deep holes, and, when the tide is nearly full, may be caught with a stout minnow-line or a very delicate paternoster. He recommends, when fishing for them in the docks, gentles, cod-baits, and blood-worms, fresh raw shrimps, or small pieces of raw lobster or crab. Some people in his day considered that the whitebait was the fry of the smelt, their habits and localities being the same.

Of the genus *Coregonus*, we have the GWYNIAD (*Coregonus clupeoides*); the VENDACE (*Coregonus vendesius*), said to have been introduced into Loch Maben, Dumfriesshire, by Mary Queen of Scots, probably with about as much foundation as the introduction of the grayling by the monks; and the POLLAN (*Coregonus pollan*).

The Gwyniad, called the Schelly in Cumberland and Powan in Scotland, and, according to Tennant, identical with the Ferar (*C. fera*) of the Swiss lakes, is silvery-blue in colour, lightest on the sides and below; dark greyish-blue fins; irides silver; pupils blue.

Fin-rays: dorsal, 13–15; pectoral, 17; ventral, 11–12; anal, 13–16; caudal, 19.

The Vendace is found chiefly in Loch Maben and the surrounding lakes, is gregarious in its habits, exceedingly delicate, and it is said not to survive if once taken out of the water, although immediately returned. Its food appears to consist of minute entomostraca. It is never taken by a bait.

Fin-rays: dorsal, 11; pectoral, 15–16; ventral, 11; anal, 13–14; caudal, 19.

The colour is greenish-blue along the back and upper half of the body; the sides silvery, with a slight golden hue; the abdomen silver-white; eyes yellow; pupils blue; dorsal fin greenish-brown; lower fins bluish-white. There appears to be seven rows of scales between the lateral line and the insertion of the ventral fin, while in the gwyniad there are eight rows, and in the pollan nine.

THE SMELT.

The Pollan is found in Ireland, chiefly in Loughs Earne, Neagh, and Derg. They are chiefly caught in nets, packed, and sent off to Liverpool and Manchester, where they are considered great delicacies. The colour is bluish-black, with silvery sides and dark fins.

Fin-rays: dorsal, 13–14; pectoral, 15–16; ventral, 11–12; anal, 12–14; caudal, 23.

Some authors have considered that these two latter are only varieties of *C. clupeoides*.

CHAPTER XV.

THE PIKE.

THE family of *Esocidæ* is the tenth of the order PHYSOSTOMI, and is characterised by the "body covered with scales; barbels, none; margin of the upper jaw formed by the intermaxillaries mesially, and by the maxillaries laterally. Adipose fin, none. The dorsal fin belongs to the caudal portion of the vertebral column. Stomach without blind sac; pyloric appendages, none. Pseudo branchiæ glandular, hidden; air-bladder simple; gill-opening very wide" (Günther).

"This family," says Günther, "includes one genus only, *Esox*, the 'Pikes,' inhabitants of the fresh waters of the temperate parts of Europe, Asia, and America. The European species, *E. lucius*, inhabits all three continents; but the North American waters harbour five, or perhaps more, other species, of which the 'Muskellunge' or 'Muskinonge' (*E. estor*), of the great lakes, attains the same large size as the common pike. The other species are generally called 'Pickerell' in the United States."

Yarrell gives the generic characters of the pike, *E. lucius*, as: "Head depressed, large, long, oblong, blunt; jaws, palatine bones and vomer, furnished with teeth of various sizes; body elongated, rounded on the back; sides compressed, covered with scales; dorsal fin placed very far back over the anal fin."

The colour of the head and upper part of the back dusky olive-brown, becoming lighter and mottled with green and yellow on the sides, passing into silvery-white on the belly. Pectoral and ventral fins pale brown; dorsal, anal, and caudal fins darker brown, mottled with white, yellow, and dark green; irides yellow.

The fin-rays: dorsal, 19; pectoral, 14; ventral, 10; anal, 17; caudal, 19.

Odd notions were held about the pike in early days. The eighth chapter of the "Compleat Angler," ed. 1653, commences thus:—" It is not to be doubted but that the Luce or Pickrell or Pike breeds by spawning, and yet Gesner says 'that some of them breed where none ever was, out of a weed called Pikrell weed and other glutinous matter; which, with the help of the sun's heat, proves in some particular ponds apted by nature for it to become pikes.'"

Moses Browne writes in one of his piscatory eclogues:—

"Say, canst thou tell how worms of moisture breed,
Or pike are gender'd of the Pickrell weed?"

As the spawn is deposited among the leaves of aquatic plants in March or early in April, one can easily understand how these early writers were misled.

Pike, in spawning-time, generally frequent the ditches and smaller stagnant waters near the main river in pairs, and there perform the necessary functions for reproducing their species; and it has been asserted by some that the female after this is over not infrequently devours her husband.

Pike commence to breed when about three years old.

The pike is certainly the most voracious of all our freshwater fishes, and therefore its presence in a trout-stream is much to be dreaded; for it appears that a small trout is a *persona grata* to this wolf of the waters.

Day makes the remark (which we can ourselves endorse) that "pike in trout-preserves not only diminish the inhabitants, but scare them to that extent that they become timorous of feeding, and frequently occasion a great falling-off in the general condition of the trout."

We noticed that in a part of the Itchen, where a few years ago there were a great number of large pike, no trout would rise in the day-time; but as soon as these were destroyed by netting and other means, the trout began to feed as usual.

These voracious fish have no hesitation in eating each other. Fishing one day near St. Albans, we rose and hooked a small jack about eight inches long. In pulling him across the river, another jack much bigger made a rush and swallowed our little friend, and after a fight we landed them both. The larger one, weighing 4 lbs., was not hooked, and shook himself free when he found himself in the landing-net. Another time, near Leatherhead, a good-sized jack was caught spinning, and as he was being landed a large pike rushed at him, took him across the back, and made off with him. After a good fight both the fish were brought to bank and landed, the larger fish weighing over 16 lbs.

Trout are often taken by jack, but last year we witnessed a case of retaliation. Strolling down a river, we noticed in the middle of a shallow something bobbing up and down, and on getting opposite to it we saw it was the tail and part of the body of a good-sized fish. But what was making it bob so? It was too much in the middle of the stream to take an accurate survey, but on wading in below, and cautiously making our way up, we were astonished to find that this proceeded from the mouth of a large fish, which was doing all it could to swallow it down; and on looking carefully we found it to be a large trout of at least 5 or 6 lbs., evidently half-suffocated. We tried to put our landing-net under it, but it was too cautious. Still, it only moved on a few yards, and again began the wobbling. Again we tried, but without success; and this went on for fully a quarter of an hour. However, in the end, after three or four endeavours, the trout began to think we were trying to play him a trick, and with an effort he disgorged the fish and went off. On securing the dead one, to our surprise we found it was a jack of at least 2 lbs., the head and shoulders actually in process of being digested.

Buckland gives the following curious circumstance :—

"Mr. Francis Crump of Killin sent me in April 1870 a box containing two pike, the two fish weighing together 19 lbs., exactly in the same position as when gaffed by the boatmen on Loch Tay. He says: 'We saw a considerable commotion in the water, and on approaching to discover the cause, the fish appeared to be fighting, and merely sank a short distance below the surface. The gaff penetrated both their heads. They both lived for some hours after they were in the boat. You will observe that the head of one fish, weighing perhaps 9 lbs., is tightly inserted up to the termination of its gill and part of the first lower fin in the mouth and throat of the larger one.' In all probability, instead of being fighting, the larger fish had tried to devour the smaller."

Pike prefer deep pools where there are abundance of aquatic plants, particularly those of the broad-leaved kind; and although those caught in rivers are the best as to their edible qualities, yet they appear to grow bigger in still waters than in streams. That they do grow to a great size and live to a great age cannot be doubted; but whether we must believe in the recorded weights and age of former days remains somewhat doubtful. The largest pike of modern times are the two reported from Ireland— one taken in August 1830 in the County Clare, said to weigh 78 lbs., and one taken at Portumna, about 1823-4, said to weigh 92 lbs. Buckland states that when Whittlesea Mere was drained a pike was found weighing 52 lbs.

The famous pike of Hailbrun, with a brass ring attached to it, with the inscription that it was put into the lake by Frederick II. in 1280, and was, therefore, two hundred and seventy-seven years old when captured in 1497, and weighed 350 lbs.—we will leave our readers to believe or not, as it pleases them.

It is said of Abbot, Archbishop of Canterbury, that he owed his position to this circumstance:—"His mother, the wife of a weaver living at Guildford, during her pregnancy dreamed that if she could eat a pike her child would be a son, and arrive at great preferment. The pike came miraculously to hand, for she caught it while dipping her pail into the river. The story of the dream was circulated. The child was a son, was befriended and put to school, and eventually became Primate of all England."

It has been generally supposed that the pike is an imported fish, and came to be cultivated in England about the time of Henry VIII., but that they existed here in Edward III.'s time is certain. Chaucer, in his Prologue to his "Canterbury Tales," says:—

"Full many a fair partrich hadde he in mewe,
And many a Breme and many a luce in stew."

An ancient MS. exists, written about 1250, in which *Lupus aquaticas sive Luceos* is amongst the fish which the fishmongers were to have in their shops.

Three lucies were the arms of the Lucy family in the time of Henry III. See note in Pickering's edition of "Walton," p. 206.

Pike generally go by the name of jack till they are twenty-four inches long. The old name was Lucy or Luce. There are many provincial names—Gullet in Northumberland; in Cambridgeshire a large pike is called a Haked, in Scotland a Gedd; in France it is *Brochet*, *Lanceron*, and *Becquet*.

Holinshed, "Chronicles of England," vol. i., says: "Also how the Pike as he ageth receiveth divers names, as from a Fire to a Gilthed to a Pod, from a Pod to a Jacke, from a Jacke to a Pickerell, from a Pickerell to a Pike, and last of all to a Luce."

The term "pike" is supposed to come from the Saxon word *piik*, sharp-pointed.

THE PIKE.

Old Barker, a rare good fisherman, in his "Art of Angling," gives the following for killing a pike:—

> "A rod twelve feet long, and a ring of wire,
> A winder and barrel, will help thy desire
> In killing a pike; but the forked stick,
> With a slit and a bladder, and that other fine trick
> Which our artists call snap, with a goose or a duck,
> Will kill two for one, if you have any luck.
> The gentry of Shropshire do merrily smile
> To see a goose and a belt the fish to beguile.
> When pike suns himself, and a frogging doth go,
> The two-inched hook is better, I know,
> Than the ord'nary snaring. But still I must cry,
> When the pike is at home, mind the cookery."

CHAPTER XVI.

THE EELS.

THE family *Murænidæ*, to which the Eels belong, is the thirty-first of the order PHYSOSTOMI.

Gunther says: "The eels are spread over almost all fresh-waters and seas of the temperate and tropical zones; some descend to the greatest depths of the oceans. Fossil remains of this family are very numerous."

The genus *Anguilla* is the only one found in our fresh-waters. It is thus defined by Day: "Gill-openings of moderate extent, situated near the base of the pectoral fins; upper jaw not projecting beyond the lower; teeth small and in bands; dorsal fin commencing at some distance behind the nape; pectorals present; small scales imbedded in the skin."

THE EEL.

Three species of eels are described by Yarrell as indigenous to Great Britain—the SHARP-NOSED (*Anguilla acutirostris*), the BROAD-NOSED (*A. latirostris*), and the SNIG (*A. mediorostris*).

Day considers these differences arise chiefly from sex or are dependent on sterility, and classes them under the name *Anguilla vulgaris*. The Grig or Glut Eel (*A. latirostris*) is mostly found in brackish water near the sea, seldom going up the rivers any distance beyond the tides. This, he states, is the male; the females being the sharp-snouted eel (*A. acutirostris*), which are taken in large numbers on their way to the sea in autumn for breeding purposes.

There is another form, very broad-snouted, which appear to be barren females, or sterile for a time. They are found in fresh-waters, and seem to be very fierce and voracious; occasionally some of these are found migrating at the annual breeding-time towards the sea.

Much has been written about the eels' migration.

Eels descend the streams and rivers to the sea for spawning purposes. When they arrive at the tidal waters they continue to descend during the ebb, but cease during the flow of the tide. This migration takes place at different periods. In Norfolk it commences in July and ends in November. In the Severn it appears to commence in October and November with the first rise of the waters; in the Kennet in August, September, and November, some descend as early as April; in the Yare and Waveny the eels come down in large solid balls from one to two feet in diameter, heads inside and tails out, and these living balls roll down the river and plump into the nets with such force as to carry them away.

This has also been seen during the migration in Ireland. It is asserted that the large eels never return to the fresh water after spawning. Buckland's opinion is that they do return, but at the same time they are not recognised, inasmuch as traps are set only for eels on their downward march. There are no traps to catch them wholesale on their upward march. The down-parents, moreover, go in large numbers; the up-parent eels, "I fancy, go singly."

At what age eels begin to breed is a problem; it is possible they remain unprolific for some years, and when ready migrate to the sea. Most writers on eels have stated that these fish breed only in brackish waters of the estuaries of our rivers, but many good observers declare that some eels do breed in fresh water, in ponds and lakes which have no connection with rivers that run to the sea. But all ponds and lakes have their outfalls, which by some means or other communicate with rivers the waters of which eventually get to the sea. That young eels migrate in vast numbers every spring from the estuaries there can be no doubt, and when very minute, will run up every watercourse, however small, and so reach the head-waters. When eels migrate down to the sea they will use any ditch or watercourse for their purpose, even travel long distances over wet meadows. Seeing small eels in freshwaters is no proof of their having been bred there.

The late Dr. Francis Day read a paper on the Mode of Propagation of the Common Eel at the annual meeting of the Cotteswold Naturalist Club, May 1886, in which he says: "For the generation of eels, it would seem, so far as we are at present aware, that the presence of salt water is a necessity; for it has been observed that when these fish leave rivers and brackish waters for the sea, their productive organs have scarcely begun to develop. But their maturing in the sea must be rapid, because in five or six weeks they have arrived at a breeding condition. This rapidity of maturing in the breeding organs would seem to be a cause of extreme exhaustion; consequently, after the breeding season is over they die, similarly to lampreys and several other piscine forms; and this furnishes the explanation why subsequent to this period old eels are not observed reascending rivers." This, however, is not to be taken as proved. We should much like to know if dead river eels have ever been found in large quantities either in the sea or in the estuaries. Buckland's hypothesis, as stated above, is much more likely to be the case.

Eels were by the earlier naturalists supposed to be viviparous and hermaphrodite, but such is not the case. Eels breed like other fish; both sexes are distinct, and the female deposits its spawn in the brackish waters of the estuaries. Gyrski was the first to discover, in 1873, the milt organs in the male eel, which lie under the two fatty folds which are found in all eels, and appear as a very narrow light band on either side of the vertebral column, and are so glass-like and transparent as to be difficult to make out except when held in an oblique direction towards the sun; and Day remarks that for the proper propagation

THE EELS.

of the species, both the male and female eel must enter salt water to develop properly the generative organs. The female organs of generation appear as frill-like bands, rather broad, extending from the liver as far as the vent, which bands contain a large quantity of fat.

Day supposes there are females which may be either permanently or temporarily sterile, and which have the ovaries in a very anomalous condition. Instead of the fatty frilled band there is only a thin frothy-looking band, destitute of fat, and having very few folds, often transparent as glass. These eels, he says, have generally, but not invariably, been found to be very broad-headed ones, possessing very small eyes. Those found in brackish water-marshes are much sought after by epicures as having a delicious flavour.

Eels are found in almost all our rivers, lakes, and ponds. They prefer clear, pure water with a muddy bottom, are very tenacious of life, yet are easily affected by sudden change of temperature. The white-bellied or silver eel is the most esteemed for the table.

Eels are certainly able to travel over moist fields from one pond to another, which may probably account for eels being found in ponds and pits, which they could not have reached in any other way. They are very susceptible of cold, and are found at times in the winter huddled together in the mud.

The young eels ascend from the sea in the spring in enormous numbers. They are called Elvers, a corruption of Eel*fare, fare* being a Saxon word signifying to go, to pass, or to travel. They mount all obstacles to gain the upper part of the rivers, and these elvers are caught in great numbers, thrown into a tub of salt, then boiled and pressed into cakes, cut in slices and fried, and are very good eating.

Mr. Sealy, of Bridgewater, proved that it took 14,087 of elvers to make a pound-cake, or upwards of three millions to the ton. What a serious injury must arise from making these cakes! In one day in 1886 three tons of elvers were sent from the Gloucester district to make into elver-cakes. The consumption of eels in London is over 1500 tons a year, of which about 1000 tons, of the value of about £80,000, come from Holland.

Mr. Willis Bund, writing of the migration of these fish, says: "The first that ascend the river are called elvers; they come in March and April. Then come a larger and darker form, from six to ten inches in length, called *Elverboults*. The next variety, from six to ten inches, are termed *Snigs;* when still larger are called *Puntcheon Eels*. Eels four or five to the pound go by the name of *Stick Eels*, and if of half a pound, as *Shutlings;* all above this weight as eels. The first eels that descend in June are 'stick eels;' the smaller form only migrate up-stream. In August and September the stick eels or shutlings arrive, and during September and October the large eels descend seawards. The foregoing are irrespective of forms which always remain in the river, and are called Glut Eels. Their heads are very large, and they are of a dark colour."

In early times eels were supposed to be generated by quite a different mode to all other fish. Aristotle supposed they sprung from mud; Pliny thought they arose from fragments of the skin of their parents rubbed off against the rocks; and Helmont says: "May dew placed between two pieces of turf has the same effect;" whilst others give horsehair, particularly that from the tail of a stallion, as a certain recipe for forming eels.

An eel is at once disabled by a strong blow on the extreme end of the body, which, either by paralysing the

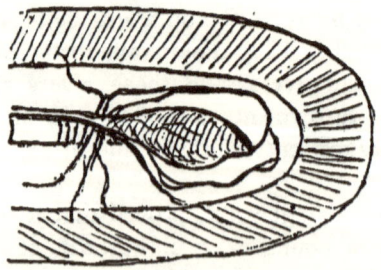

THE CAUDAL HEART.

spinal chord, or it may be by crushing that curious pulsating organ which is placed near the tail, and which

Müller calls the caudal heart, renders the fish unable to use the powerful muscles of the tail.

Much has yet to be learned of the natural history of this fish, particularly as to its migration, mode of propagating its species, and the localities in which that process takes place, whether only in brackish or salt water, or in both fresh and salt water. Up to the present time the evidence is against the fish breeding in fresh water. Much more positive information is required on this particular point.

THE LAMPREYS.

In the third sub-class, *Cyclostomata*, or circular-mouthed fishes having a cartilaginous skeleton, without ribs or real jaws, are found—

The LAMPREYS, which belong to the first family of this third sub-class—the *Petromyzontidæ*—characterised by the body being eel-shaped and naked, and subject to metamorphosis.

THE LAMPREY.

Günther ("Study of Fishes," p. 170) says: "Changes amounting to metamorphosis have been hitherto observed in *Petromyzon* only. In the larval condition (*Ammocætes*) the head is very small, and the toothless buccal cavity is surrounded by a semi-circular upper lip; the eyes are extremely small, hidden in a shallow groove, and the vertical fins form a continuous fringe. In the course of three or four years the teeth are developed, and the mouth changes into a perfect suctorial organ; the eyes grow, and the dorsal fin is divided into two divisions."

Petromyzon (Stone-Sucker) has two dorsal fins, the posterior joined with the caudal; two teeth in the jaws, placed close together, the lingual teeth serrated.

The SEA-LAMPREY (*Petromyzon marinus*) ascends the rivers to spawn in the spring, and is caught in considerable numbers in the estuary of the Severn and the river itself. They were formerly, and still are occasionally, found in the Thames, also in some of the rivers in Scotland and Ireland. Lampreys are said to die after spawning.

The LAMPERN or SILVER LAMPREY (*Petromyzon fluviatilis*), also known as Lampron or Lamper Eel, Nine Eyes, Seven Holes, Spanker Eel, Say-Nay, &c., is very common in many of the rivers in England, Scotland, and Ireland, more particularly in the Thames, the Severn, the Kennet, the Dee, the Trent, the Tweed, and many others. At Tewkesbury, on the Severn, there is a large manufactory for potting lamperns.

The lampern spawns from April to June. Sometimes they go in pairs, sometimes in numbers, to the breeding-ground, while the male and female act together in removing stones in order to prepare the spot. We have seen this process going on in the Kennet, and the size of the stones moved by these small creatures is extraordinary.

Their food consists of insects, worms, and the flesh of dead fish. Day says they affix themselves to living fish, among which they are accused of doing much injury. Thomson says he received one which was taken adhering to a large trout.

The adult fish is usually from twelve to fifteen inches in length, the body slender, cylindrical for two-thirds of its length, then compressed to the end of the tail; head rounded, with a single aperture on the crown. The lips surrounding the mouth have a continuous row of small points on their margin; the back is furnished with two rather elongated dorsal fins.

CHAPTER XVII.

MOLLUSCS, CRUSTACEANS, AND ANNELIDÆ (EARTH-WORMS).

WE cannot roam about the banks of rivers or ponds without constantly coming in contact with certain animals commonly called slugs and snails, the first being without covering, and the second forming its own habitation in the shape of a shell. They both belong to the *Gasteropodous Mollusca*. The slugs, family *Limacidæ*, are closely allied to the *Helicidæ*, but have no external shell; those usually met with are—

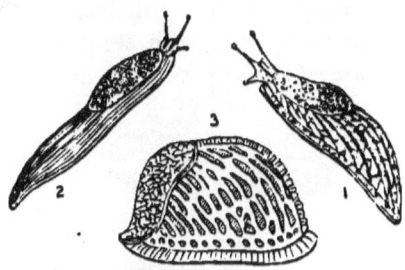

The GREY SLUG (*Limax agrestis*), Fig. 1, of a white ash-colour.

The GREAT GREY SLUG (*Limax maximus*), the largest of our British species.

The BLACK SLUG (*Limax ater*), Figs. 2, 3, commonly known as the Black Snail, is mentioned by Izaak Walton as a bait for chub, "with his belly slit to show the white." Fig. 3 shows the same at rest.

The RED SLUG (*Agrion agrestis*). This mollusc is by some supposed to be carnivorous. It is very fond of certain fungi and decayed vegetable matter.

Of the Fresh-Water Snails, there are five genera which inhabit our fresh-waters, and are excellent scavengers, feeding on confervoids and decaying vegetable matter; they are hermaphrodite, depositing their eggs on stones and aquatic plants, enveloped in masses of a slimy substance. They are also of special value as food for trout.

The genus *Planorbis* (flat coil) is at once distinguished by the shell being flat and concave, resembling an ammonite. The animal emits a kind of purple blood, and most of its vital organs lie on the left side of the body.

There are nine British species, some of them very small, all inhabiting ponds, rivers, lakes, marshes, and wet ditches; those commonly met with are—

Planorbis lineatus (Fig. 3), often found on duck-weed.

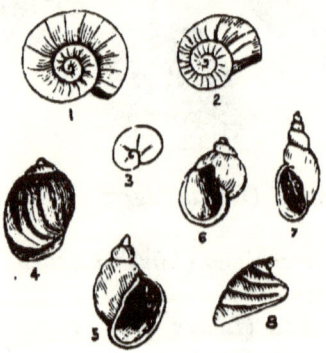

Planorbis marginatus (Fig. 1), often found feeding on decayed leaves of the yellow iris (*Iris pseudo acorus*), and other water-plants. *Planorbis carinatus.* (Fig. 2).

The next two genera are distinguished from *Planorbis* by having *spiral conical shells*, the genus *Physa* having the shell conic oval, or *oblong spiral sinistral*, *i.e.*, turning from right to left; the shells having a remarkable polished appearance.

These little molluscs are mostly found in clear shallow running water.

There are two British species. The most common is *Physa fontinalis* (Fig. 6)—often found attached to water-cresses and other aquatic plants in running streams, and also in canals and ponds; some of the shells are milk-white.

THE MOLLUSCS.

In the genus *Lymnæa* the shell is conic oval, or *elongated*, the spiral turning from left to right or dextral.

There are eight British species. Those commonly met with are *Lymnæa peregra* (Fig. 4) or the Wanderer, *Lymnæa stagnalis* (Fig. 5), and *Lymnæa palustris* (Fig. 7). There are fourteen varieties of this mollusc.

The genus *Ancylus* (Fig. 8) is at once distinguished by its shell being hood-shaped. There are two species: *Ancylus fluviatilis*—shell semi-oval or curved towards the front like a Phrygian cap, not glossy, of a yellowish-grey or horn colour—found abundantly on stones and rocks in shallow rivers and streams; and *Ancylus lacustris*, found chiefly on the under side of the leaves of water-lilies and other aquatic plants; shell oblong, obliquely twisted to the left, glossy, greyish horn-colour.

In the genus *Neritina* there is only one species, *Neritina fluviatilis*—shell semi-globose, cornea above slightly compressed towards the spire, almost concave below, yellowish-brown with zig-zag bands. It is found in rivers and streams and ponds having a gravelly bottom. Both trout and grayling are very fond of this mollusc.

Of the Land Snails, there are two which often come under our notice out of the vast number comprised in the three families *Limacidæ*, *Testacidæ*, and *Helicidæ*.

The *Helix aspersa* has a reddish-brown shell with a single white band—commonly known as the Garden Snail.

HELIX ASPERSA.

It is found in woods and gardens, and is one of the shells forming the *kitchen midden* of the thrush, this bird being extremely fond of the animal itself, as well as of the *Helix memoralis*, which is also found in woods, hedges, gardens, and is very common. The shell of *H. aspersa* is glossy, generally yellow, brown, pink, or white. There are four varieties of this snail.

In Cowper's minor poems there is a short one on the snail, which well illustrates the habits of our common garden species:—

" To grass, or leaf, or fruit, or wall,
　The snail sticks close, nor fears to fall,
　As if he grew there, house and all
　　　　　　Together.

　Within that house secure he hides,
　When danger imminent betides
　Of storm, or other harm besides,
　　　　　　Of weather.

　Give but his horns the slightest touch,
　His self-collecting power is such,
　Hs shrinks into his house with much
　　　　　　Displeasure.

　Where'er he dwells, he dwells alone,
　Except himself has chattels none,
　Well satisfied to be his own
　　　　　　Whole treasure.

　Thus hermit-like his life he leads,
　Nor partner of his banquet needs,
　And if he meets one, only feeds
　　　　　　The faster.

　Who seeks him must be worse than blind
　(He and his house are so combined),
　If, finding it, he fails to find
　　　　　　Its master."

There are twenty-four British species of *Helix*, large and small, some very minute.

Of the Crustaceæ, two will occasionally come under notice—

1. The CRAYFISH (*Astacus fluviatilis*) or the River Lobster (*Astakos* being the name by which the Greeks called the

lobster) is found in many of our rivers. It is of a dull-greenish or brownish colour, pale-yellow on the under side, with a little red on the limbs. The female has a much broader tail than the male. It eats every imaginable aquatic insect or animal, from the larvæ of insects to the water-rat. It also eats calcareous plants, as the stoneworts, and even the weakly members of its own family are devoured. For a detailed account of the natural history of this crustacean see Professor Huxley's most interesting volume, "The Crayfish."

Owing to some unknown cause, the crayfish has entirely

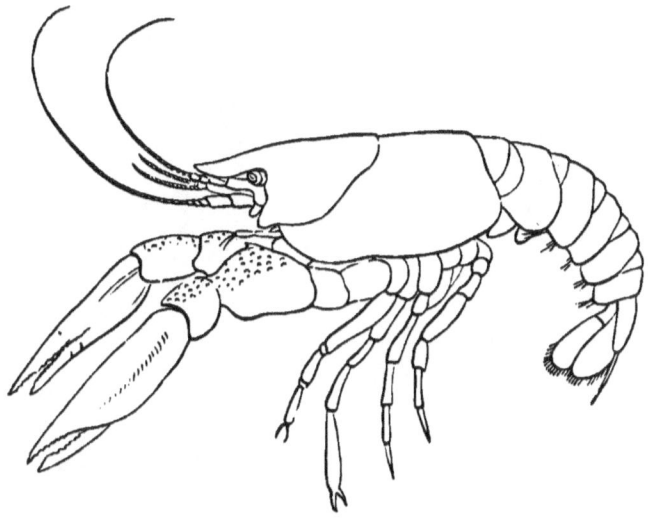

1. THE CRAYFISH.

died out from the upper part of the river Kennet, and consequently the trout have lost a most important food-supply; and it is possible that the redness of the flesh for which the trout in this river were noted, and which is not now so universal, was due in a great measure to this crustacean, to the young of which, trout are extremely partial. May not the cause arise from the absence in the water of ingredients which were necessary for the formation of the shell?

2. The FRESH-WATER SHRIMP (*Gammarus pulex*) is extremely common in all springs and rivers, particularly where decaying vegetable matter has accumulated. It generally keeps near the bottom, and swims on its side with a kind of jerking motion, and feeds on dead fishes or any other decaying matter. In some parts of the Kennet this crustacean is to be found in great numbers.

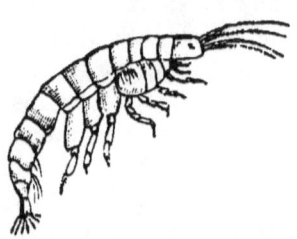

2. THE FRESH-WATER SHRIMP.

CHAPTER XVIII.

THE INSECTS.

A CAREFUL study and close observation of the various forms of insect life found either in, or on, or about the water is of the greatest interest to those who frequent the river-side for the purpose of fly-fishing, as it is chiefly by the successful imitation of many of those beautiful and minute denizens of the air and water that they are enabled to capture their prey. The difficulty of making an identical imitation of many of these is extremely great—it even may be doubted if an exact copy of the prototype could ever be produced.

In almost all imitations of the natural insect the number of legs, and sometimes of the wings, are not in accordance with nature, and it must be borne in mind that trout are daily becoming more educated in their perceptive faculties. There is still, however, some chance for the angler.

All fly-fishers know that trout and grayling will at times not only refuse the natural fly floating over them, and its artificial similar, but will at once greedily take some fancy fly. Do they take it for some juicy morsel, or do they at times follow the example of their cousins, the salmon and sea-trout, and take flies which have none of the characters and colours of those usually found on the river? That trout do take all kinds of things into their mouths as they float over them there is no question. We were watching some big trout at the tail of a mill-stream, and whilst doing so we lighted our pipe and threw the fusee into the water; as it sank, a great trout (at least 3 lbs. in weight) at once took it into his mouth. He soon rejected it, but had there been a hook attached to the fusee the trout would have had a bad time of it. We were one day fishing on the Kennet; there was a great rise of the pale-blue dun, which the trout were taking eagerly. They would not look at the artificial similar, but at once, and greedily, took a small imitation of the March brown (which was not on the water); and, without moving ten yards, we landed five brace of fish, most of which we had seen rising at the blue dun.

Insects pass through four metamorphoses during their life—

First, the *Egg state*, which is a non-active state.

Second, the *Larva state*. This term of larva, from the Latin word signifying a mask, was given by Linnæus, because he considered that the real insect when in this form was masked; this name is also given to caterpillars, grubs, and maggots, which latter are generally called "gentils" or "gentles," a very ancient name. Tusser says:—

> "Rewerd not thy sheep when ye take off his cote
> With twitches and patches as brode as a grote;
> Let not such ungentleness happen to thine,
> Least fly with her *gentils* do make it to pine."

During this second state the larvæ are very voracious, cast their skins several times, and live for a long or short period—some only for a few days, others for weeks, months, or even years. When they cease eating they generally fix themselves in a secure place, their skin separates once more and discloses an oblong body, and they now enter the third (*Pupa* or *Chrysalis*) state, which may be divided into—

(*a.*) That which differs but slightly from the larva state, but which has a kind of rudimentary wing. These are not only extremely active, but very voracious, as the dragon-flies, &c.

(*b.*) That which is motionless and quite inactive, often surrounded with a hard case, as the cocoon or chrysalis of butterflies, moths, &c.

This state may continue, as said above, for long or short periods, and then the fourth or last state is entered—viz.,

THE INSECTS.

the *Imago* or perfect state—when the rudimentary wings become perfect, and the insect is able to take flight, and carry out the process of reproduction, &c. The *Ephemeridæ* appear to be an exception, having a state (*Sub-Imago*) between the third and fourth.

The different orders of insects which will come under our notice are—

Coleoptera . . .	Beetles, &c.
Neuroptera . . .	Nerve-winged, as Dragon-Flies, May-Flies, Duns.
Trichoptera . . .	Insects produced from Case-Worms, Caddis, as the *Phryganidæ*.
Orthoptera . . .	Grasshoppers, &c.
Hemiptera . . .	Bugs, &c.
Hymenoptera . .	Bees, Wasps, Ants, &c.
Diptera	Two-winged insects, House-Flies, Gnats, &c.
Lepidoptera . .	Moths, Butterflies, &c.

We all know those little black, metallic, shiny fellows which are constantly gyrating in pools where the water is somewhat quiescent, and which are known commonly by the name of Whirligigs; the French call them Tourniquets, the scientific name being GYRINUS NATATOR. They appear

GYRINUS NATATOR.

to be always arranging a kind of skating quadrille, in and out and round about, and again joining in the centre. Just make your presence known by disturbing their pastime with the top of your rod. Ho, presto! Where are they? disappeared, for they are wonderful divers. Be careful, if you wish to examine them with your pocket-glass—and they are worth examining—to have your hand gloved, and even then you will not like it, for they emit a most unpleasant odour not easily forgotten. You will find that they have the antennæ very short; the two forelegs long and stretching forwards, other legs short and broad; the eyes are divided by horny processes, making one into two, as it were. They live in the mud in winter, deposit their eggs on the leaves of aquatic plants, and feed on small aquatic insects.

The GREAT WATER-BEETLE (*Dytiscus marginalis;* from the Greek word *dytes*, a diver) is another familiar object coming at times to the surface of the water to breathe. He is a regular cannibal, with an enormous appetite, and will devour anything that comes in his way. It is generally supposed that the fish will not touch these beetles, but we are not so sure. At night they often leave the water and fly

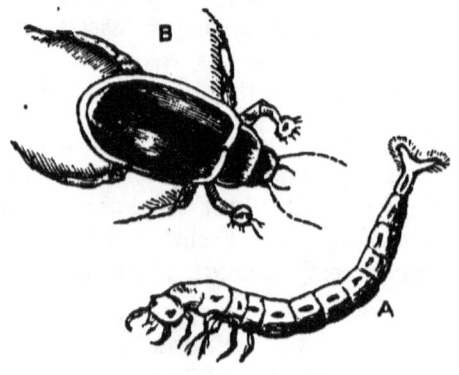

A. THE LARVA.   B. THE PERFECT INSECT.

about, and perhaps as they return to their usual element, when their aerial flight is over, as they touch the surface of the water they may be gobbled up by the great trout, who are lurking about in those hours of darkness. They are about $1\frac{1}{4}$ inch in length, of a dark olive-colour. The throat and outer edge of the wing-case has a yellow margin. There are numerous species, varying in size.

Another very common object is the WATER-BOATMAN or BOAT-FLY (*Noctonecta glaucus*; order, *Hemiptera;* sub-order, *Heteroptera*). It, too, is a voracious eater, and is extremely partial to tadpoles, and also its own species. It

THE WATER-BUG.

has the peculiarity of always swimming on its back, using as oars its hinder legs, which have a very powerful propelling power. It is about half an inch in length, generally of a greenish tinge, but varies much in colour, some being quite black. It very seldom flies, but it can do so, and very well. When diving it carries down a globule of air, for the purpose of respiration, in a hollow between the folded wings.

THE WATER-BOATMAN.

One of this family (*Corixa*) is, however, often found in the stomachs of trout and grayling; but this differs somewhat in its mode of progression, as it does not swim on its back.

In the still eddies of streams, as well as on quiet pools and slow-running rivers and mill-heads, we constantly see that long-legged, brownish-looking, stick-like insect, the WATER-BUG (*Ranatra linearis;* order, *Hemiptera*).

It seems to glide or skate, as it were, along the surface, and after three or four long slides comes to a full stop. It has two pairs of wings and three pairs of long legs.

In Gilbert White's time it was supposed that this species was viviparous, and he gives a curious account of the mode of propagation; but the eggs have long been known to entymologists, and are deposited, or rather inserted, into the stems of aquatic plants, usually the club-rush, in which it is

THE WATER-BUG.

so deeply embedded by the long ovipositor of the insect as to be entirely hidden, the bristles alone projecting. These bristles, by preventing the edges of the plant stem from uniting, secure an exit for the larva when hatched. The females of this insect are much larger than the males.

Many of us, no doubt, have come across that curious, black-looking insect, the DEVIL'S COACH-HORSE or COCK-

TAIL BEETLE (*Gœrius olens*), easily recognised by the peculiar way it has of sticking up its tail, or rather the hinder part of its body, when disturbed. It emits a very offensive odour, and feeds on carrion; also on living insects, as the following interesting account by that pleasant writer, the "Amateur Angler," of an encounter with this insect, shows. He says: "As I was hurrying the other day, on fish destruction bent, I was suddenly stopped by a curious adventure on the pavement. One insect had got another by the throat, and was struggling to carry him along from one side of the path to the other; it was like a puny infant of four trying to carry a fat thirty-pound baby. I touched the murderer with the point of my stick, a little black wretch, something over an inch in length; he dropped his

THE DEVIL'S COACH-HORSE.

THE SAME, SHOWING THE WINGS.

victim, a grasshopper quite as big as himself, turned, cocked up his tail, which seemed to have eyes in it, stared at the stick as much as to say, 'Who are you? Do that again if you dare.' Then he turned again, seized his wriggling prey, and was making off with it. I again gave him a poke. 'Heyho!' thinks he, 'there's something up; I must be off.' He started, leaving the dying hopper on the pavement. When he had got a yard away I retired three or four yards; he stopped, cocked up his perky tail. 'Heyho!' says he again, 'the coast is clear; I'm not going to leave that fat grasshopper for nothing.' Then he began circling round and round, gradually nearing, till he was within a few inches, then he stopped, cocked up his tail as usual, and pretended to look surprised, as if he had never seen a grasshopper before in his life. 'Hello!' says he, 'why, that must be a hopper.' Then he made a rush at him, seized him by the throat, and was making off, when I again interrupted him. Then he threw him down in disgust, and started off at full speed towards the hedge. I did not think it necessary to arrest him, but I considered it only an act of mercy to end what seemed to me the very painful, though it may have been pleasurable, but certainly dying struggles of the poor hopper, by smashing the little life that was left in him into nothing with my boot. The performer in this small tragedy used to be called, when I was a country boy, '*The Devil's Coach-Horse*,' and I know of no other or more appropriate name for him. Scientifically I believe he is known as *Staphylinus (Ocypus) olens*.

"I wonder how he first got hold of the springy and active hopper: probably caught him napping, and then his first operation was to ham-string him with his powerful mandibles; the rest was easy enough."

As regards the scientific name, Westwood, "Entomologist's Text-book," gives that of *Gærius olens*; Newman, "History of Insects," *Staphylinus*. "Those beetles," he says, "are distinguished from all others by their square, short, forewings, naked body, elongate form, and disgusting manner of turning up the tail like a scorpion. It devours all putrefying substances, as well as living insects."

Stephens places *Ocypus* as a distinct genus, and as being much smaller; less rapacious than *Gærius*, and inhabiting sandy heaths, and having curved simple mandibles alike in both, instead of being dissimilar, as in *Gærius*.

We must not omit to mention one of the *Coleoptera*, which, coming out only in the evening, often startles us with its loud hum as it rushes by to bury itself in some mass of dung or carrion. This is the COMMON DOR-BEETLE (*Geotropus stercorarius*), the shardborne beetle of Shakespeare:—

"Ere the bat hath flown
His cloistered flight; ere to black Hecate's summons
The shardborne beetle, with his drowsy hums,
Hath rung night's yawning peal!"
—*Macbeth*, Act iii. sc. 2.

Gilbert White says:—

> "While deepening shades obscure the fall of day
> To yonder bench, leaf-shelter'd, let us stray,
> Till blending objects fail the swimming sight,
> And all the fading landscape sinks in night,
> To hear the drowsy dorr come brushing by
> With buzzing wing, or the shrill cricket cry."

This beetle is a blackish-blue in colour, with a beautiful blue metallic lustre on its under surface, about three-quarters of an inch in length; being a carrion eater, it is anything but pleasant to the olfactory organs.

THE DOR-BEETLE.

There are between three and four thousand British species of beetles, and as yet only a few are recognised as lures for trout. Ronalds only gives three; but there must be many which fall from the branches of trees and shrubs overhanging the water, or from the stems of aquatic plants, or blown on to the surface by the wind, which are taken by all kinds of fish.

The word *Coleoptera* is derived from two Greek words— κολος, a sheath, and πτερον, a wing; and this order is more especially distinguished by having the wings enclosed in a pair of scaly cases, called elytra, the wings themselves being membranous in character, and carefully folded under their horny cases.

The PEACOCK-FLY of Ronalds is the *Lathobrium quadratum;* family, *Staphylinidæ*. Another of this family, *Oxiporous Rufus*—red, with the head and elytra black, frequenting brambles and whitethorn bushes—must be often blown on to the water. This appears to us to be the natural simile of the Red Tag, so very killing in some waters, particularly for grayling.

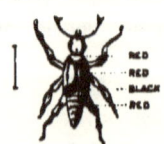

OXIPOROUS RUFUS.
(The line shows the natural size).

THE INSECTS.

The *Pæderus Riparius*, of the family *Stenidæ* — head black; neck, thorax, and abdomen orange, or orange-rufous; elytra dark green; found in moist places, on the banks of rivers, and under stones — is, we think, the *Orange Tag* of the fly-makers.

PŒDERUS RIPARIUS.

The FERN-FLY of Ronalds (*Telephorus livida;* family' *Telephoridæ*), known also as the Soldier-Fly, Soldier Palmer; has red elytra, and is very killing at certain times. The Soldier-Fly is also known to some as the *Red Spinner*—a misnomer, as the latter is really the imago state of one of the *Ephemeridæ*. The Sailor-Fly (*Telephorus fuscus*) has dark brownish-blue elytra instead of red.

The MARLOW BUZZ or COCH-A-BONDHU, SHORN-FLY (*Chrysomela populi;* family, *Chrysomelidæ*), is a very common beetle, particularly in June, and feeds upon the poplar-leaves; it has a black thorax, with rufous elytra. There are twenty-six species of the genus *Chrysomela;* one, *C. Fulgida*, four or five lines in length, of a bright golden-green, is found in marshy places and on river-banks. Is this the *Green Insect*?

THE MARLOW BUZZ.
1, Larva; 2, Perfect Insect.

The MAY-BUG of Norfolk, BRACKEN-CLOCK of anglers, is *Anisoptia saturalis;* family, *Melalonthidæ*. Shiny-black, with green lines, sparingly clothed with long hairs; elytra dull ochreous; shoulders and margins black; tarsi ferruginous; appears in May, June, and July. The imitation of this fly is given in Hofland's " Manual."

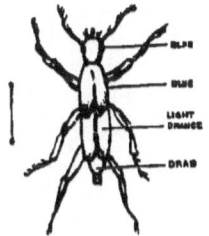

The *Blethisa Multipunctata* — family, *Carabidæ*—with its bronzy, shining elytra, about six lines in length, found on the borders of lakes and rivers, amongst reeds, &c., must often become a tempting ormsel to the trout.

ODOCANTHA MELANUSA.

The *Odocantha Melanusa*, with its blue head and thorax and testaceous elytra, inhabiting reeds and marshy ground, would be almost the natural simile to the *Macaw Tag*.

Many others might be enumerated; but sufficient have been given to show that the smaller beetles found in the vicinity of rivers must often become food for fish.

The reader is referred to the coloured representations of the above in Curtis's "Coleoptera" and Spry and Schuckard's "British Coleoptera."

CHAPTER XIX.

THE order *Neuroptera* (nerved wing), from the Greek words, *neuron*, a nerve, and *pteron*, a wing, include a great number of interesting insects.

Westwood says: "This order is chiefly distinguished by the structure of the wings, which are naked—that is, not enclosed by elytra or tegmina—and furnished with a great number of nerves, which gives the appearance under the microscope of a piece of beautiful network."

"These nervures," says Newman ("History of Insects"), "divide the membrane into small compartments called cells. These cells are very constant in their form and proportion in the same genus of insects, consequently their variations distinguish one genus from another. The strong nervure which runs along the upper edge of each wing, either on the extreme edge or just below it, is called the *costal nervure*, and the portion, if any, above the nervure the *costal cell*, or if divided by minor nerves, *costal cells*. The incrassated portion of this nervure, which is frequently observable at about two-thirds of the distance from the body towards the apex of the wing, is called the *stigma*. The cells immediately beyond the stigma, towards the extreme point of the wing, are called *marginal cells*. The wings of all insects present a somewhat triangular figure."

WING OF A NEUROPTEROUS INSECT ENLARGED.

1, Costal Nervure; 2, Stigma; 3, the Marginal Cells; 4, the Sub-marginal Cells.

"The form of the wing in *Neuroptera* is generally somewhat elongated, as well as the body, and particularly the abdomen. The head is often large, the compound eyes very large; some have only simple or stemmatic eyes. The habits in the larva state are predaceous, which continues in some genera to the pupa and perfect states. The power of flight in many is very great. The larvæ and pupæ are often

THE GREAT DRAGON-FLY.

aquatic. The females have no sting, and only a few have an ovipositor. The metamorphosis is complete in some, incomplete in others."

THE INSECTS.   311

Dragon-Flies, May-Flies, Scorpion-Flies, &c., belong to this order.

The typical *Neuroptera*, according to the Linnæan order, are the DRAGON-FLIES (*Odontata;* family, *Libellulidæ*); highly organised with regard to their powerful mandibulate mouths and densely reticulated wings. There are three genera, *Agrion, Libellula,* and *Æschna,* and fourteen British species divided into two sections.

These insects often attract our attention when by the water-side by the curious rustling of the wings of the larger species, as they cross and recross our path on a hot

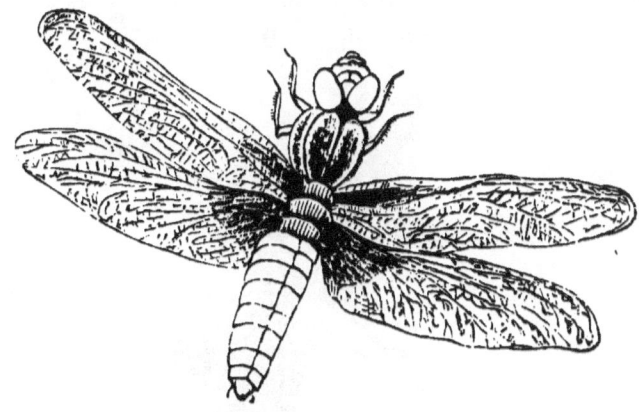

THE DRAGON-FLY OR HORSE-SINGER.

summer's day, or by the beautiful colours of the smaller kinds as they settle on the flags or other plants.

Those generally met with are—

1. The GREAT DRAGON-FLY (*Libellula grandis, Æschna grandis*), a most beautiful insect with a reddish-brown body with white spots, the wings with a marginal spot.

2. The DRAGON-FLY or HORSE-SINGER (*Libellula depressa* —Donovan). Eyes brown; thorax greenish, with two yellow transverse bands; a large dark spot at the base of each wing, and a small dark spot on the anterior margin. Donovan says it is a perfect vulture among lepidopterous and other defenceless insects, destroying more for its sport than for its voracious appetite.

3 and 4. *Libellula puella* and *Libellula cæruleus*. The *Libellula puella* has the body of a greenish-blue; wings of equal length, with a cloudy, brown spot in the middle. The length of the insect is about two inches. It is often to be seen sporting on the waters in May and June. There is a variety with the body red, with yellow and black lines; the wings clear, with marginal spots.

The dragon-flies are very remarkable for the rapidity and power of motion. Kirby, on "Motion of Insects," says: "Their four wings, which are nearly equal in size, are a complete and beautiful piece of network, resembling the finest lace, the meshes of which are usually filled by a pure, transparent, glassy membrane. In two of

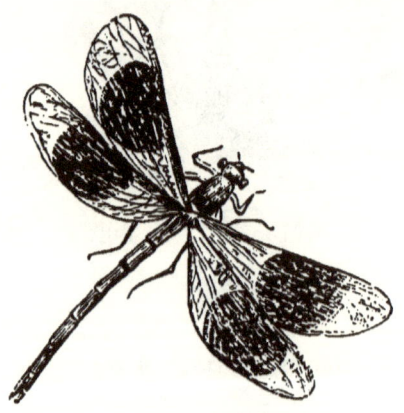

THE LIBELLULA PUELLA.

the genera belonging to this tribe the wings, when the animal is at rest, are always expanded, so that they can take flight in an instant, no previous unfolding of these organs being necessary. In *Agrion*, the other genus of the tribe, the wings when they repose are not expanded. I have observed of these insects that, without turning, they can fly in all directions—backwards, and to the right and left, as well as forwards. This ability to fly all ways without having to turn must be very useful to them when pursued by birds. Leeuwenhoek once saw a swallow chasing an insect of this tribe in a menagerie about one hundred feet long; the little insect flew with such astonishing rapidity to the right, to the left, and in all directions, that this bird of rapid wing and ready evolution was unable to overtake and entrap it, the insect eluding every attempt, and being generally six feet before it. The species of the genus *Agrion* cut the air with less velocity; but so rapid is the motion of their wings that they become quite invisible."

Dragon-flies are most abundant near water, as they deposit their eggs in that element, the larvæ and pupæ being entirely aquatic. They are great devourers of all other aquatic insects.

That the larvæ of dragon-flies are very deleterious to the fry of fish the following will testify :—

"In the Hungarian *Rovotani Lapok* of December last, L. Bizo states that the larvæ of some of the *Libellulæ*, species not determined, have made such ravages in the piscicultural establishment of Count Palffy at Izomolany, that in a pond in which 50,000 young fish were placed in the spring of 1884 only fifty-four could be found the following September; but there was a large quantity of the *Libellulæ* referred to" (*Zoologist*, April 1885).

The metamorphosis from the pupa to the imago is extremely interesting. When the final transformation is about to take place, the pupa crawls out of the water on the nearest object to it, either a rush or stem of a plant or a stick, fixes itself by its legs, which are furnished with hooks, and the skin then splitting up the back, the imago or perfect insect is released. The body and wings, which are still folded up, are, however, quite soft and moist. The wings become rapidly developed, and it dries both its body and wings in the sun, and soon takes its departure, roaming in all directions.

Sometimes these insects appear in enormous quantities. Kirby mentions that Meinechen once saw, on a clear day, such a cloud of dragon-flies as almost to conceal the sun. At times, in Germany, cloud-like swarms of the *L. depressa* were seen at Weimar and other places extending over a very large district.

"The larvæ of certain dragon-flies (*Æschna* and *Libellula*)' will afford you the most amusement by their motions. These larvæ commonly swim very little, being generally found walking at the bottom on aquatic plants; when necessary, however, they can swim well, though in a singular manner. If you see one swimming you will find that the body is pushed forward by strokes, between which an interval takes place. The legs are not employed in producing this progressive motion, for they are then applied close to the sides of the trunk in a state of perfect inaction, but it is effected by a strong ejection of water from the

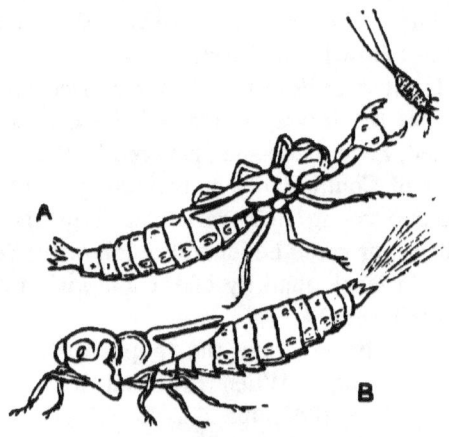

PUPA OF THE DRAGON-FLY.
A—The mask extended. B—Mask closed, and discharging a current of water.

posterior orifice (see fig. B above). This is produced by drawing the water in and then expelling it. As the larva between every stroke of its internal piston has to draw in a fresh supply of water, an interval must of course take place between the strokes" (Kirby and Spence, "Motions of Insects").

The author of the article "Insects" in the "Encyclopædia Britannica" adopts metamorphosis as the basis of classification of the *Neuroptera*. He says: "The stumbling-block of all systems has been the Linnæan order *Neuroptera*, inasmuch as its members combine the characters of most of the other orders, and ingenious

THE INSECTS.   315

American writers have attempted to overcome this difficulty by considering it a collection of 'synthetic types.' In adopting metamorphosis as a basis of classification, we prefer to take another course, and to follow Erichson, who (in 1839) boldly transferred all those *Neuroptera* with incomplete metamorphoses to the *Orthoptera* as a sub-order, although, in dealing with the *Neuroptera* in the light of a specialist, division into several orders appears the more natural course."

The orders, according to this author, with the *metamorphosis complete* are—*Hymenoptera* (bees, &c.) ; *Diptera* (flies, gnats, &c.) ; *Lepidoptera* (butterflies, moths, &c.) ; *Neuroptera*, divided into two sub-orders—*Trichoptera* (sedge-flies, &c.) and *Planipennia* (saw-flies, &c.).

With the *metamorphosis incomplete* are placed the order *Orthoptera*, divided into two sub-orders—*Pseudo-Neuroptera* (the *Ephemeridæ*, &c.) and *Genuina*—and the order *Hemiptera*, divided into two sub-orders—*Heteroptera* and *Hemiptera* (grasshoppers, &c.).

We propose in describing these to alter the classification in which they are thus placed, and to take them, as better suited for our purpose, in the following order:—

1, The *Neuroptera*, with its two sub-orders, *Trichoptera* and *Planipennia;* 2, the *Orthoptera*, with its sub-order *Pseudo-Neuroptera;* 3, the *Hemiptera*; 4, *Hymenoptera;* 5, *Diptera;* and lastly, the *Lepidoptera.*

The *Neuroptera* (with *metamorphosis* complete) are described as having four membranous wings, for the most part densely reticulate, more or less clothed with hairs, but without true scales ; very frequently the hairs are on the neuration only ; mouth mandibulate. Pupa has its members free. This order is divided into two subdivisions—*Trichoptera* and *Planipennia.*

The *Trichoptera* (*hairy-winged*) or *Caddis-Flies* "form a very natural and sharply defined group, distinguished by their rudimentary mouth parts, with the exception of two pairs of palpi,[1] which are strongly developed, the maxillary

[1] The palpus is a slender appendage, somewhat similar in structure to the antennæ, only shorter, and composed of fewer joints, varying

pair being the longer, and often with the greater number of joints. The antennæ setaceous; wings with simple neuration and but few transverse nervules; ordinarily covered with hairs, which sometimes simulate scales. Larvæ (known as Caddis-Worms) with well-developed thoracic legs and anal crotchets, but without prolegs, living in tubes covered with extraneous materials; pupa lying free in the case, or occasionally in a special cocoon; only active just before its metamorphosis; habits, with one or two exceptions, aquatic."

This order is divided into seven families, distinguished chiefly according to the structure of the maxillary palpi—the *Phryganidæ, Lymnophilidæ, Sericostomidæ, Leptoceridæ, Hydropssychidæ, Rhyacophilidæ*, and *Hydrophilidæ;* of which the first, the *Phryganidæ*, is what we have chiefly to do with. In the two latter the larvæ inhabit fixed cases; in the others the cases are free, and carried about by the inmates.

The *Phryganidæ* are generally found in the neighbour-

DIFFERENT FORMS OF CADDIS CASES WITH THE SILK GRATING.

hood of water. The larvæ live upon aquatic insects and leaves of water-plants, and reside in the water in cases made up of all kinds of substances—shells, bits of stick, sand, &c.—which the insect collects as a house to live in before taking on the pupa state. Westwood, in describing

from one to six articulations. It is attached close to the base of the outer lobe of each maxilla.

this state, calls it incomplete—that is, having all the limbs distinct, but folded upon the breast, the head being furnished with a pair of curved mandibles, which appear to have no other purpose than of making a passage through an open-work grating of silk which the larva had formed at the opening of its case previous to assuming the pupa state. "When, therefore, the time arrives for the insect to quit its watery abode and assume the winged state, it is endowed with powers of motion far greater than is possessed by any other incomplete pupa, so that it is enabled, not only in the first place to cut through the grating of silk, but to creep out of its case, and then, rising to the surface, it crawls up some plant, where it throws off its outer skin."

Kirby and Spence, in the "Introduction to Entomology —Habitation of Insects," give a most interesting account of these larvæ:—

"The larvæ of the various *Phryganeæ*, a tribe of four-winged insects, which an ordinary observer would call moths, but which are even of a distinct order (*Trichoptera*), not having their wings covered by the scales which adorn the *Lepidopterous* race. If you are desirous of examining the insects to which I am alluding, you have only to place yourself by the side of a clear and shallow pool of water, and you cannot fail to observe at the bottom little oblong moving masses, resembling pieces of straw, wood, or even stone. These are the larvæ in question, well known to fishermen by the title of caddis-worms, and which, if you take them out of the water, you will observe to inhabit cases of a very singular formation. Of the larva itself, which somewhat resembles the caterpillars of many *Lepidoptera*, nothing is to be seen but the head and six legs, by means of which it moves itself in the water and drags after it the case, in which the rest of the body is enclosed, and into which on any alarm it wholly retires.

"The construction of these habitations is very various— some elect four or five pieces of the leaves of grass, which they glue together in a shapely polygonal case; others employ portions of stems of rushes placed side by side, forming an elegant fluted cylinder; some arrange round them pieces of leaves, like a spirally rolled ribbon; others enclose themselves in a mass of leaves of any aquatic plants; others, again, form their abodes of minute pieces of wood; others construct houses which may be called alive, forming them of shells of various aquatic snails of different kinds and sizes. The case of *Leptocerus bimaculatus*, formed of mixture of sand and mud, is pyriform, and has its end curiously stopped by a plate formed of grains of sand, with a central aperture. . . . Their mode of proceeding in arranging the necessary compensating balance is still more curious. Not having the power of swimming, but only walking at the bottom of the water, it is of great importance that its house should be of a specific gravity, so nearly that of the element in which it resides, and it is essential that it should be equally *ballasted* in every part as to be readily movable in every position. Under these circumstances our caddis-worms evince their proficiency in hydrostatics, selecting the most suitable substances; and if the cell be too heavy, gluing to it a bit of leaf or straw, or if too light, a shell or piece of gravel. It is from the necessity of regulating the specific gravity, that to the cases, formed with the greatest regularity, we often see attached a seemingly superfluous piece of wood, leaf, or the like."

A very favourite fly on some rivers in April and May is the GRANNOM, one of this family. Ronalds states that it is of the genus *Tinodes;* family, *Phryganidæ*. According to Curtis, the grannom is *Linnephilus nervosus;* family, *Phryganea*. It is very local in its distribution.

The RED or CINNAMON SEDGE is *Phryganea varia* (Curtis). The SILVER SEDGE *Limnephilus flavicornis* (Curtis), or *Phryganea rhombica* (Donovan).

PHRYGANEA VARIA.

There are forty-eight British species of this genus, varying from each other. One species, *Phryganea grandis*, is very large, appearing about the same time as the May-Fly.

The SILVER HORNS of Ronalds (the dark silver twist of fly-makers, so well known as a first-rate grayling fly) is

THE INSECTS.

Leptocerus niger or *longicornis*. There are twenty-four British species of *Leptoceridæ*, differing slightly from each other.

Ronalds gives a figure of the SAND-FLY, and places it in the genus *Phryganea;* but the term "Sand-Fly" is very confusing. In the north, according to the author of "The Angler and the Loop-Line," the sand-fly of that part of England is the Gravel-Bed or Spider-Fly of the south. Order, *Diptera;* family, *Tipulidæ*.

Somewhat the same confusion exists as to the STONE-FLIES. The Stone-Fly of Ronalds is *Perla bicaudata;* family, *Perlidæ*. Newman, "History of British Insects," places all the stone-flies or *Phraganites* in the family *Phryganidæ;* Westwood places them in the genus *Phryganea;* and Mr. Cholmondeley-Pennell, "Angler Naturalist," p. 90, gives two figures to represent two states of the Stone-Fly, *Phryganea grandis* (common caddis-worm of anglers).

"The *Planipennia*, or true *Neuroptera* (according to modern ideas), have strongly developed mandibulate mouths, and for the most part monoliform or filiform, often claval antennæ. The wings, ordinarily densely reticulate, with very numerous transverse nervules, the membranes hairless, or nearly so; the pupa ordinarily in a cocoon, active just before its transformation. *Planipennia* are subdivided into *Panorpites* (Scorpion-Flies), *Sialidæ*, and *Megaloptera*." The *Panorpites* (Scorpion-Flies) are found plentifully in the woods and hedges.

The *Sialidæ* have the wings depressed and reticulated, with strong nervures, which frequently vary on the two sides in the same insect. There is only one indigenous species, which frequents the borders of rivers and streams, *Sialis luteus*, well known as the ALDER-FLY, or ORL-FLY. The body is a deep black; wings fuscous, with black nervures. The female lays a great quantity of brown conical eggs on the leaves of aquatic plants. The larvæ are very active and swim well.

THE ALDER-FLY.

The insects included in the family *Perlidæ* have the wings incumbent during repose; the reticulations vary. The head large; ocelli three; abdomen frequently with two articulate setæ; tarsi always simple. They frequent damp places. Those indigenous to this country are divided into five genera—

1, *Perla;* 2, *Isogenus;* 3, *Chloroperla;* 4, *Nemoura;* 5, *Leuctra.*

In the genus *Perla* (Geoffroy) the insects are usually of large size, and the sexes are very dissimilar, the males having the wings much abbreviated. Their legs are also longer, and their general habits very dissimilar to the females, which have large expansive wings and shorter legs, the palpi in both sexes of different length, the maxillary ones much longer.

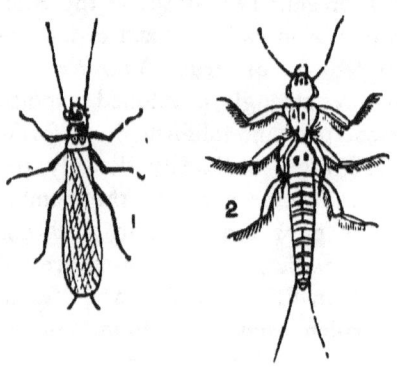

PERLA MARGINATA.
1, Perfect Insect; 2, Larva.

Perla marginata is the Stone-Fly of the north of England and Scotland, where it abounds in June and beginning of July. Described as follows :—

" Fuscous; antennæ as long as the body; head moderate, black, varied with yellow, with some glossy elevated spots, a longitudinal dorsal channel having an elevated line on each side; abdomen more or less ochreous or yellowish on its sides and towards the apex; setæ as long as the antennæ; legs yellowish; wings fuscescent with darker nervures; pupa pale, fuscous, with dusky spots" (Stephens).

THE INSECTS.

Ronalds gives *Perla bicaudata* as the stone-fly, which is found in June on the banks of the Thames and elsewhere. "Fuscous; head with irregular longitudinal tawny or orange streak behind; abdomen ochreous beneath; setæ nearly as long as the body; antennæ as long as setæ; wings fuscescent, with darker nervures; a tawny or orange longitudinal streak along the body" (Stephens). Newman places it in the next genus, *Isogenus*, on account of the sexes being similar in habit, by both possessing ample wings, and calls it *Isogenus nubecula*, the stone-fly: darkbrown; anterior wings hyaline, slightly tinged with brown, and having a little oval cloud of a darker brown on the costal margin, situated about one-third of the distance from the tip towards the body; posterior wings beautifully hyaline iridescent; legs pale-brown.

Mr. Newman says that this species is abundant in the neighbourhood of running water in Herefordshire, Worcestershire, Nottinghamshire, &c., and is a favourite food of trout and grayling.

The YELLOW SALLY, *Perla lutea* of Ronalds, is placed by

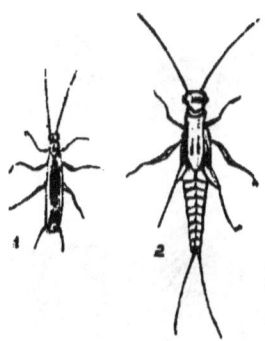

THE YELLOW SALLY.
1, Perfect insect; 2, Larva magnified.

Mr. Newman in the genus *Chloroperla*—*Chloroperla lutea*; the *Perla flava* of Pictet.

The various WILLOW-FLIES belong to the genus *Nemoura*, which Stephens divides into fifteen species; fourteen of them have a X-like plexus of nervures towards the apex

x

of the wings. Stephens says *Nemoura variegata* is destitute of this; Pictet, however, gives the X plexus in this species.

The Willow-Fly, with dark purple body, may be *Nemoura*

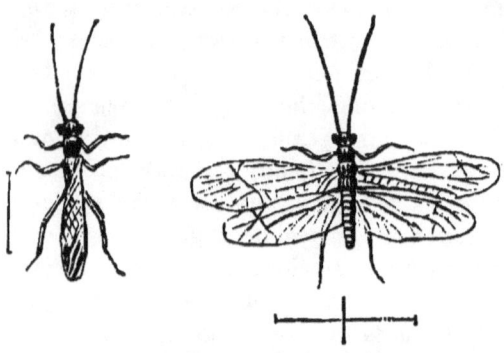

NEMOURA VARIEGATA.

variegata of Pictet. The Common Willow-Fly, *Nemoura annulata*, of Stephens.

Pictet's "Familles des Perlides," with its beautiful plates, should be consulted by those interested in this family.

In the order *Orthoptera* are the *Locustidæ*, one species of which, the GRASSHOPPER (*Stenothrus viridulus*), with short

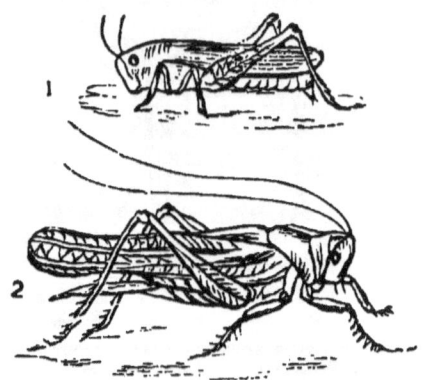

1. THE COMMON GRASSHOPPER.   2. THE GREAT GREEN GRASSHOPPER.

antennæ, is much used as a lure for grayling, trout, and other fish, and must not be confounded with the Great Green Grasshopper with *long antennæ*, sometimes found in moist meadows and other places — *Gryllus viridissimus*, *Locusta viridissimus*.

THE INSECTS.

The Grasshoppers, so well known to all fishermen, and, as one author remarks, associated with the brightest of green pastures and sunshiny days, are divided into two families—the *Locustidæ*, of which we believe there are many species, and the *Grillidæ*, of which the Common Cricket (*Gryllus domesticus*) and the equally familiar Cockroach (*Blatta*) are examples.

The so-called singing of these insects is caused by the rubbing together of the talc-like substance at the base of the wing-cases, and this only occurs in the males.

The American editor of "The Complete Angler" says, in a note to Chap. V.: "Dipping for trout with the grasshopper, the grass-beetle, the cricket, &c., over a running stream is the nearest imitation, as it was the original, of fly-fishing. Many an 'attic minstrel' have I, when a boy, made to seduce the shy, speckled, shining beauties from their haunts; but no artificial imitation have I ever succeeded with, though the very counterfeit of life—the trout will not take them."

CHAPTER XX.

The family of the *Ephemeridæ* comprises various genera and species, and offers peculiar attractions to the fly-fisher. Stephens ("Illustration of British Entomology," vol. vi., 1835) gives the British genera in the following order :—

1, *Ephemera*, with 16 species ; 2, *Cœnis*, 7 species ; 3, *Baëtis*, 18 species ; 4, *Cloëon*, 8 species.

Pictet (" Hist. Nat. des Insectes Neuroptères," 1843) gives seven genera of the *Ephemeridæ*, five of which are British, viz. :—

1, *Ephemera*, 2 species ; 2, *Baëtis*, 5 species ; 3, *Potomanthus*, 3 species ; 4, *Cloëon*, 3 species ; 5, *Cœnis*, 2 species.

Eaton (" A Revisional Monograph of Recent Ephemeridæ or May-flies," 1883-4-5) gives twelve genera, viz. :—

1, *Ephemera*, 3 species ; 2, *Potomanthus*, 1 specie ; 3, *Leptophlebia*, 2 species ; 4, *Harbrophlebia*, 1 specie ; 5, *Ephemerella*, 1 specie ; 6, *Cœnis*, 3 species ; 7, *Baëtis*, 9 species ; 8, *Centrophelum*, 2 species ; 9, *Cloëon*, 3 species ; 10, *Siphlurus*, 2 species ; 11, *Nithrogenia*, 1 specie ; 12, *Heplagenia*, 2 species.

Most of these new species of Eaton's have been recognised both by Stephens and Pictet, only under different names, and placed generally in *Baëtis* and *Cloëon*.

The author of the article " *Ephemeridæ* in the " Encyclopædia Britannica " (vol. viii.) places this family in the *Pseudo-Neuroptera*, a sub-order of the *Orthoptera*, having the metamorphosis *incomplete*. He says : " The *Ephemeridæ*, a remarkable group of pseudo-neuropterous insects deriving the name from εφημερος, in allusion to the very short lives of the winged insects. In some species it is possible that they have scarcely more than one day's existence ; but others are far longer lived, though the extreme limit is probably rarely more than a week. These insects are all aquatic in their preparatory state. The eggs are dropped into the water by the female in large masses, and it is probably several months before the larvæ are excluded. The sub-aquatic condition lasts a long time in the genus *Cloëon*, a small and delicate species, Sir John Lubbock having proved it to extend over six months; but in the larger and more robust genera (*e.g.*, *Planigenia*) there appears reason to believe that the greater part of three years is occupied in the preparatory conditions.

"The larva is elongate, the head rather large, furnished at first with fine, simple eyes of nearly equal size; but as it increases in size the homologues of the faceted eyes of the imago become larger, whereas those equivalent to the ocelli remain small. The antennæ are long and thread-like, composed at first of a few joints, but the number of these latter apparently increase at each moult; mouth-parts well developed; powerful mandibles; three distinct and large thoracic segments; the proto-thorax narrower than the others; legs much shorter and stouter than in the winged insects; abdomen, ten segments, the tenth furnished with long and slender multi-articulate tails (setæ), which appear to be only two in number at first, but an intermediate one gradually develops itself (though the latter is often lost in the winged insect). Respiration by external gills, placed along both sides of the dorsum of the abdomen and hinder segments of the thorax. According to Lubbock and Joly, the very young larvæ have no breathing organs, respiration being effected through the skin. Lubbock traced at least twenty moults in *Cloëon*. At about the tenth, rudiments of the wing-cases began to appear. These gradually become larger, and when so the creature may be said to have entered its nymph stage; but there is no condition analogous to the pupa state of insects with complete metamorphoses. There appear to be three or four different modes of life in these larvæ—some are fossorial, and form tubes in the mud or clay in which they live; others are found on or beneath stones; while others, again, swim and crawl freely among water-plants. Probably some are carnivorous, either attacking other larvæ or subsisting on more minute forms of animal life; others perhaps feed on vegetable matter of a low type, such as diatoms.

"The winged insect differs considerably in form from its aquatic state. The head is smaller, often occupied almost entirely above in the male by the very large eyes, which in some species are curiously double in that sex, one portion being pillared, and forming what is termed a turban. The mouth-parts are aborted, for the creature is now incapable of taking nutriment, either solid or fluid. The antennæ are mere short bristles; the proto-thorax is much narrowed, the other segments greatly enlarged; the legs long and slender; tarsi four or five jointed; wings carried erect—anterior pair large, with numerous nervures and abundant transverse reticulations; the posterior pair very much smaller, often lanceolate, and frequently absolutely wanting. Abdomen consists of ten segments; at the end are either two or three long multi-articulate tails (setæ). In the male the ninth joint bears forcipated appendages; in the female the oviducts terminate at the junction of the seventh and eighth ventral segments. The sexual act takes place in the air, and is of very short duration; but is apparently repeated several times.

"The number of described species is not less than 200, spread over many genera."

From the earliest times attention has been drawn to the enormous abundance of some species of this family in certain localities. Scopoli, writing more than a century ago, speaks of them as so abundant in Carniola, that in June twenty cartloads were carried away for manure. *Polymitarcys virgo*, which though not found in England, occurs in many parts of Europe (and is common in Paris), emerges from the water soon after sunset, and continues for several hours in such myriads as to resemble snow-showers, putting out lights and causing inconvenience to man and annoyance to horses by entering the nostrils. In other parts of the world they have been recorded in

THE INSECTS.

multitudes that obscured passers-by on the other side of the street.

The *Ephemeridæ* belong to a very ancient type of insects, and their fossil imprints are common, occurring even in the Carboniferous system.

Stephens says: "The insects of this family are remarkable for undergoing a quadruple metamorphosis, as in addition to the ordinary states of egg, larva, pupa, and imago, there is an intermediate one to the last two, inasmuch as a sort of representative of the imago is produced after the pupa, but which has to undergo a further ecdysis or shedding of its skin before the true insect appears. In this state the insect is capable of flying, the true wings being encased in a delicate membrane, which is cast off very expeditiously when sufficiently mature. In this operation a slit is made in the back, through which the insect forces itself, and gradually withdraws its body, limbs, and wings, leaving its exudium exactly corresponding with its previous form, excepting that portion which enveloped the wings, and which is rolled up in a mass on each side of the thorax." He is of opinion that the popular idea that May-flies are strictly ephemeral—that is, live but for twenty-four hours—is fallacious in most respects. Owing to the atrophy of the mouth-organs and the condition of the abdominal cavity, the perfect insect cannot eat, and is only born for the purpose of propagating its species, and he thinks certainly some of the family may be able to exist for two or three days.

The Rev. A. E. Eaton, in his monograph on the *Ephemeridæ*, describes them as insects with a long, soft, ten-pointed, sessile abdomen, furnished at its hinder extremity with either two or three many-pointed setaceous or filiform tails (caudal setæ), and whose body is smooth and glabrous, head free, with atrophied mouth-organs, carinated epistoma, short subulate antennæ, composed of two short, stout joints, succeeded by a slender many-pointed setaceous awn, three ocelli and large oculi (compound eyes, always larger in the male), thorax robust, sternum well developed. The abdomen in the male is furnished with a pair of forceps placed at the extremity of its last segment; alimentary canal capacious and filled with gas. The fore-wings are usually trilateral, ample, and rounded off at the extremities; the margins unequal in extent, erect or spreading in repose, plaited lengthways, but not folded up. The hind-wings when developed are sometimes very minute—never large, generally oblong-ovate in form. The inner margin of the fore-wing and anterior margin of the hind-wing hitch together automatically to a larger or smaller extent when the wings are spread open. The legs are slender, femora strong, the fore coxæ somewhat distant from the others. The legs present great differences in their relative lengths of the several pairs, either sexual or generical. The fore-legs are always longest in the male, and generally longer than the hinder pairs—usually three pairs, with five tarsal joints.

The forceps of the male is two, three, or four-jointed, and in some genera (according to Mr. Eaton) afford good distinctive characters of species.

As regards the caudal setæ, Mr. Eaton says there is much diversity in the number and relative proportions. Often all of one length, but the middle one is often either shorter or longer than the others; often atrophied to a mere rudiment, or altogether absent. The outer setæ persistent, and, according to sex or genus, exceed many times, or fall short of, the length of the body.

Mr. Eaton says: "The popular supposition that May-flies are strictly ephemeral is fallacious in most instances. Provided the air be not too dry, the imagines of many genera can live without food several days. The diurnal *Ephemeridæ*, of which our May-fly is one, in very hot weather rest during the midday heat, generally flying in the cooler hours of sunshine, or just after sunset. Many species fly by night only, and most of them couple by night, male below, gradually sinking nearer the ground; then, when accomplished, the male rises again in the air to resume its search for other females."

The eggs in some of the short-lived species, when laid by the female, are discharged from the ovaries in egg-clusters, which rapidly burst, and the eggs sink, and lie broadcast on the bottom of the river.

The less perishable species extrude their eggs gradually, part at a time, and deposit them in one or other of the following manners :—Either the mother alights on the water at intervals to wash off the eggs that have issued from the mouths of the oviducts during her flight, or, as Mr. Eaton has himself observed, she creeps down the stem of a plant into the water, enclosed in a film of air, and deposits her eggs (the wings being collapsed so as to cover up the abdomen) under stones, &c., and on completing her labour, floats up to the surface and flies away. But at times her setæ get wet ; she is unable to extricate herself, and is drowned.

A letter in the *Field* of April 10, 1886, from Mr. W. F. Beart, of Godmanchester, corroborates these observations. He was returning from fishing in a stream in Bedfordshire in the early part of May, 1885, when he noticed near a wooden bridge a number of spinners, some on the water, some in the air. His attention was attracted to an insect sitting on a post of the bridge, quite close to the water—an olive dun. He saw it walk down the post head foremost into the water, the pressure of which appeared to double the wings flat on the back, and so disappeared. In about a minute it walked up the post again, no longer an olive dun, but a spinner. He was so much interested that he remained for an hour, and saw a number of olive duns do the same thing, some rising into the air, others floating down the stream on the surface of the water.

Mr. Eaton says : "The duration of the egg state varies with the temperature to which they are exposed"—and makes some interesting remarks as to the power of living of the *Nymphæ* in relation to the temperature of the water. He says : "Besides the influence of flood and drought, or constancy of supply, the climate of the water is largely concerned in determining the fitness or unsuitability of a particular site for particular kinds of the *Ephemeridæ*. A knowledge of the water climate needed by a species renders intelligible the limitation of its geographical and local distribution. The temperature of the ordinary land-springs in a district enables the climate of other water in that neighbourhood to be ascertained readily by comparison with it. If the water of a given site exhibits marked differences in temperature from the standard of the neighbourhood according to the season or time of day, its climate is extreme, and the site cannot be inhabited by species which require relatively cold water." Sir John Lubbock states that during the first few days after birth the young cast their skins several times, the intervals between the moultings increasing by degrees. Adolescence is evidenced by the advancement of the reproductive organs internally, and externally by the growth of rudimentary wings from the hind-quarters of the proper segment. The forceps of the male also begin to bud forth, and in their general form most nymphs nearly resemble the adult. Mr. Eaton prefers the term nymph to designate all the sub-aqueous stages in the early life of these insects, instead of that of larva or pupa, which, in earlier authors, denoted respectively the wingless and wing-budding conditions.

He says: "May-fly nymphs mostly feed upon either mud or minute aquatic vegetation, such as cover stones, and the larger plants; but (judging by their mandibles and maxillæ) some must be predacious. Many of them live in concealment in the banks or under stones in the bed of streams, rivers, and lakes; others ramble openly among water-weeds, and swim with celerity. Certain genera are restricted exclusively to large rivers; and one of these (*Palingenia*) is said to remain a nymph for three years."

PUPA (NYMPH) OF EPHEMERA VULGATA (Westwood).

How long the nymph of our Mayfly proper (*Ephemera vulgata* or *danica*) remains in the nymph state is not yet thoroughly ascertained. It often lives in the mud, and is found either close to the surface or deeper down, according to the condition of the atmosphere.[1]

---

[1] The larvæ are often found in gravelly shallows close to the surface

THE INSECTS.

In the genus *Ephemera* are our May-flies proper.

Mr. Eaton makes three species, *Ephemera vulgata*, *Ephemera danica*, and *Ephemera lineata*, to which latter he attaches the synonym *Ephemera danica, Pictet.*

This genus has the following distinctive characters:— Eyes simple on both sides, always separated by a considerable space; wings, four, with numerous transverse nervures; the posterior (small) wings about the fourth size of the anterior (large), having their nervation complete and the costal border angular. The tarsæ (terminal part of the legs) five-jointed, of which the first is very short; the abdominal segments ten in number; the setæ three, generally equal in length.

Stephens says that the eyes are rather large, ovate, united on the crown in the males, somewhat remote in the females; hence the great number (sixteen) of his species of this genus.

More recent authors have removed insects with these characters into other genera of the same family.

Stephens wrote his work on the *Mandibulata* in 1833; Pictet published his on the *Neuroptères* in 1843; the Rev. A. E. Eaton's monograph appeared in 1883.[1]

Mr. Eaton considers *Ephemera danica* to be the May-fly of anglers, the female and male of the sub-imago (*i.e.*, as the insect emerges from the water) being respectively designated by them as the Green Drake and the Bastard Drake; and he says that *E. danica* usually inhabits colder and swifter waters than *E. vulgata*.

Eaton differs somewhat from Pictet. We have placed the descriptions of the two authors in juxtaposition for the sake of comparison.

The term sub-imago denotes that stage after the wings are fully expanded, previous to the insect casting off

early in spring, and Mr. W. P. Crake records fish being caught in April with their stomachs full of May-fly larvæ, and also in August.

[1] For a more extended and scientific account of the *Ephemeridæ*, we refer our readers to the Rev. A. E. Eaton's elaborate and exhaustive monograph in the *Transactions of the Linnean Society*.

its last integument and assuming the imago or perfect state.

Pictet. EPHEMERA DANICA.—*Male Imago* (perfect insect). Head black; thorax black, brown above; mouth yellow; all the rest of the body of a pale-yellow chestnut-brown; abdomen marked with four rays above and under the lateral marks on the upper segments and six on the lower; wings transparent, colourless with brown nervures, the subcostal thicker and darker than the others; anterior legs brown, with tawny tarsi, other legs tawny; setæ pale-brown, tinged with black. Spots on the wing form an incomplete band as in *E. vulgata*, a large spot near the base of each anterior wing. Posterior wing with no centre spot, but bordered with brown.

EPHEMERA DANICA (Pictet).
E. LINEATA (Eaton).
Imago, male.

Female Imago.—Colouring generally lighter than the male; spots on the wing less distinct.

Sub-Imago.—Darker in colour; has a light spot shaped like a lance in the middle of the meso thorax.

Eaton. EPHEMERA DANICA.—*Male Imago.*—Head and thorax, upper part deep blue-black, polished; face, mouth, pale greenish-yellow; abdomen, the first four or five dorsal segments ivory-white, with a pale anereous broad triangular blotch at the base on each side, the hinder segments varied with pitch-brown, or deep-brown ochre; wings faintly grey-tinted, most distinctly so towards the anterior terminal margin and in contiguity with the black nervation; spots on the wing pitch-brown, but not so distinctly marked as in *vulgata;* legs pitch-black, hinder pair tinged olivaceous; the fore-wings darkened; setæ black; forceps sepia-brown.

Sub-Imago.—Wings at first greenish-grey, becoming greyer, variously edged with grey-black along the terminal margin; discal spots and most of the cross veinlets black;

the rest of the nervation concolorous with the wing membrane or yellowish-green.

Female Imago.—Very like the male, but with dorsal

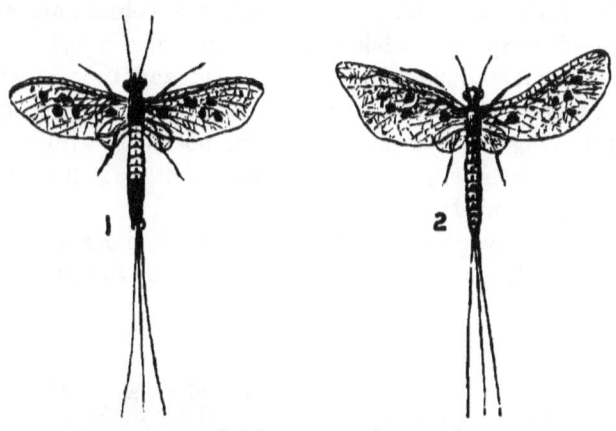

EPHEMERA DANICA.
1, Male Imago. 2, Female Imago.

markings of the pale cebreaceous abdomen better defined; fore-legs piceous, hind-legs olive-grey; wing membrane colourless, hence the narrow grey bordering of the cross veinlets is better shown than in the male; spots grey; setæ brown-black.

Pictet, 1843. EPHEMERA VULGATA. —*Male Imago.* — Head and throat peach - brown; mouth and sides of throat yellow; abdomen yellow, a long black mark on the edge of each segment; two of the same form above and two below the latter are not seen on the first segments, and become larger in the lower divisions, covering them for the most part; fore-legs black, hind-legs greyish; setæ dark-brown, ribbed with black; anterior wings tinted with brown, nervures dark brown with numerous transverse veinlets; spots on wing vary in individuals, gene-

EPHEMERA VULGATA.
Imago, male.

rally only two or three, at times running into one another; posterior wings colourless except at the extreme ends, where they are brownish. A dark spot in the centre of each wing.

The Sub-Imago only differs in that the colours are more confused. Spots on abdomen less distinct.

The female Imago is paler in head and thorax, wings much more lightly tinted with a greenish-yellow, the spots on the wing more separated. The posterior wings have the extremities slightly darker than the rest, the centre spot almost indistinct.

The Sub-Imago only differs in its paler colouring.

Eaton, 1883. EPHEMERA VULGATA.—*Male Imago.*—Head

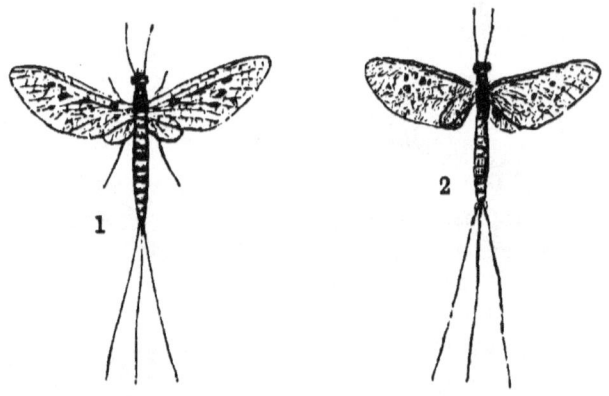

EPHEMERA VULGATA.
1, Sub-Imago, male. 2, Female Sub-Imago.

peach-brown; eyes sepia-brown; abdomen pale olivaceous, with pitch-brown markings tinged with ochre at the upper segments and in the middle line of some of the hinder segments; legs olivaceous or olive-brown; setæ and forceps pitch-brown and lutescent; wings pale greenish-grey tinted or yellowish-green, tinted with black nervations, many cross veinlets bordered with pitch-brown; spots on the wing very marked, pitch-brown.

The Sub-Imago.— Wings at first yellowish or greenish-grey, changing to ancreous, broadly tinged with black-grey along the terminal margin; discal spots fuscous, cross veinlet edged with ivory-black.

THE INSECTS.

Female.—Wings paler than the male; sometimes a spot in the middle of the posterior wing.[1]

From a number of observations made by the river-side by Mr. W. P. Crake and ourselves on the different genera and species of the *Ephemeridæ* taken in 1886 and 1887, we found that all these insects, from the time they rose from the surface of the water, remained in the sub-imago state as near as possible forty-eight hours before changing to the imago or perfect insect. In this latter state they lived from six to eight days, and in some of the genera, as *Baëtes* and *Cloëon*, as long as ten days. Whether they would thus live in a state of nature is doubtful, as some writers observe that probably the act of procreation shortens the lives of most of the *Ephemeridæ*, and, therefore, in a free state would not live so long as those kept in confinement.

The mode of transformation from the sub-imago to the imago state is accomplished in the following manner:— Somewhere about the forty-eighth hour the insect commences to flutter its wings very rapidly, so much so that the movement could with difficulty be followed. This rapid action continued from thirty to fifty seconds, then ceased, and the wings, previously upright, became perfectly flat and applied close to the body, with a very slight movement of the head and thorax. These parts split open and the perfect insect began to emerge, and rapidly divested itself of its previous covering, the whole process taking about a minute to complete. Sometimes the insect could not get the setæ, which appear folded up, clear of the old covering; and when this occurred it became very excited, fluttering the wings and bending the body backwards and forwards in its endeavours to release itself. The setæ, short in the sub-imago, become in many species of great length in the perfect state. We found also that the colour of many of these insects varied with their age. For instance, on Wednesday, April 21, 1886, ten olive duns, captured as they left the water, were put into a glass case,

[1] We have found the spots in the posterior wings of both male and female, but they are more constant in the male.

and the case exposed to light and sun. In forty-eight hours they began to change as described above. When emerged from the sub-imago to the imago state the abdomen was of a light yellow, the segments lighter still. After some hours the colour darkened, and on the 24th, *i.e.*, in three days, they had all become a dark-reddish, claret-colour, the males a deeper colour than the females. These insects lived eleven days; so that an insect caught immediately after it had assumed its imago state might easily be described as a species different to the same insect caught two or three days after.

In the May-flies proper the same process was observed; the imago or perfect insect bursting its case about the same time and in the same manner.

As regards the change from the nymph to the sub-imago state, amongst the smaller genera it is difficult to detect the exact moment of the change, but with the larger species (May-flies) the nymph rises to the surface, and it is often some little time before it can free itself from its nymph-covering (shuck), floating and fluttering along the surface of the water for a few moments before rising in the air. It is then that its life is so often cut short by the greedy trout.

The rise of the nymph of the May-fly from the bed of the river appears to be very uncertain, and from our observations does not so much depend on the state of the atmosphere as is supposed. Sometimes it would appear in considerable numbers as early as 9.30 A.M. On other days (these observations were made in 1885) it would be much later—the temperature of air and water being the same in both instances. The effect of the sun's rays on the hatch was very variable. On one cold blustering day, towards the afternoon, the rise was something extraordinary, swarms of the fly pervading air and water. On another day, with a warm sun, the hatch was but small.

May-flies remain attached to the under-surface of the leaves both of trees and bushes during the early hours of the day. But generally between two and three o'clock, whatever might be the state of the atmosphere, most of them left their resting-places and commenced their peculiar up-and-down motions, or dances in the air, the males predominating; and when the females arrived these dances greatly increased in rapidity, the union of the sexes taking place at all hours during these movements. We have often seen as many as six males attached to one female, gradually descending to the surface of the water or the ground before parting asunder, and in one instance we found a male *Danica* in union with a female *Vulgata*, with a male *Vulgata* clinging to the body. It appears to us, from carefully observing numbers of males and females, that in all probability a promiscuous intercourse between the two (so-called) species is not uncommon.

Our observations agree with Mr. Eaton's description of the *Ephemera danica*, whether it be a distinct species or a variety. We found both species in very considerable numbers, the *E. danica* being more numerous.

There were certain marks which we found pretty constant. The spots on the anterior wings in the *Danica* were as a rule five, one being close to the base of the wing. In *E. vulgata* the spots were usually three, the abdomen always as described by Eaton, the anterior segments being ivory-white.

The spot in the posterior wing absent in *Danica*, usually present in *Vulgata*, particularly in the male.

The males of both species much smaller than the females, and appearing a day or two before the females.

The sub-costal nervure, colour of the legs, and setæ varied considerably.

The male May-fly in his perfect state is seldom a prey to fish. He lives, when not resting under leaves, well above the river, dancing in the air, waiting for his wives. The swifts, swallows, chaffinches, and other birds are its enemies at that period of its existence. But when the duties are performed for which he has been brought into existence, in a short time he falls helpless and dies, often floating down the stream, then eagerly taken by the fish, and in that condition known in angling parlance as the spent gnat.

The female, when she throws off her sub-imago state, flies away to find a husband, the two combined falling gradually to the ground, or to the surface of the water; but just before reaching it the male detaches itself and mounts again in the air, while the female proceeds to deposit her eggs on the surface of the water, which sink at once to the bottom of the river. She then has to go through many dangers in her short career; she may be snapped up as she is laying her eggs either by fish or bird, or she may get her wings wet and be unable to rise, and so is hurried away by the stream, utterly helpless, and dies exhausted.

The eggs extrude from the seventh ventral segment in the female, and we observed that she sheds or lays a certain number every time she dips on the surface of the water. We caught them just after dipping, and then found none ready for immediate expulsion; and we caught them after having risen and remained in the air for a few moments, and then found a cluster of eggs half-exuded and ready for expulsion.

The May-fly season of 1887 was characterised by the fly first appearing some days later than is usual, and also in some localities by enormous numbers, which pervaded the streams and adjacent bushes and meadows. Another peculiarity was, that there was no morning hatch, although there was for the most part a bright sun, but the wind was either north-east or east. The fly generally did not hatch till after midday, and then in small quantities, the great hatch taking place from 3.30 or thereabouts, and continued till 8 or 9 P.M. On some afternoons the air appeared full of a dense mist, which was nothing but swarms of May-flies, and the accumulation of the dead was so great at some of the hatchways and elsewhere as to produce a most nauseous odour.

Mr. Eaton states "that many of the characters upon which the classification of the *Ephemeridæ* was formerly based have proved to be unsuitable for the purpose. Originally *the number of the caudal setæ* was deemed a matter of primary importance; and when forms were discovered with the median seta abbreviated, they were ranked between those with three long setæ and those with two only. Subsequently, in addition to the setæ, *the number of the wings* was employed as a leading clue to the arrangement of the genera. But it is now well known that these criteria are serviceable, at the most, for nothing more than the distinguishing of genera very intimately related to each other, belonging to various subordinate alliances comprised within the family; while one of them (the number of setæ) is not always available even for this purpose, varying, as it does in some forms, with the sex or with the individual." A variety of distinctions of modern writers, such as the cross veinlets in the wings, the *structure of the tracheal branches of the nymph*, and other modifications of organs, can hardly fail, he thinks, to be unnatural and arbitrary. "And it is only," he says, "by taking cognisance of points of difference and agreement in many details, in the anatomy and the mode of development, and the habit of leading representatives of the various alliances of genera, at different periods of their lives, before and after their exclusion from the egg, that the mutual affinities of the several associations of genera to one another can be demonstrated adequately. Until such comparisons can be, and shall have been, carred out, the whole question of their arrangement can only be dealt with in a tentative and experimental manner; and it will be fortunate if error be avoided in the necessary grouping of genera into provincial alliances of apparently kindred forms, preparatory to the study of their affinities. It is more easy to demonstrate defects in proposed methods of classification than to devise a trustworthy system in their stead; and possibly extended observation in the future may eventually show that some of the bases of arrangement adopted in this present work are mere temporary expedients worthy of mention in this paragraph."

Mr. F. M. Halford has made and carefully recorded a great number of observations on the *Ephemeridæ*, which we have every reason to hope will eventually do much to clear away the obscurity which surrounds the various genera and species of this family of insects.

The genus *Baëtis*, in which will be found many of the smaller *Ephemeridæ*, which are at times so prevalent on most of our chalk-streams, is defined as follows:—

Stephens:—Head large and transverse; eyes large, united on the crown in males, remote in females; thorax ovate, stout; wings, four, the anterior long, narrow, obtuse, considerably reticulated; posterior small, somewhat ovate; abdomen moderately long and tapering, *two* filaments at base; anterior legs long.

He divides his eighteen species into—A, wings distinctly and rather thickly reticulated; B, wings faintly reticulated.

Pictet:—Eyes *simple* in both sexes, but *much larger in the male*, in which they are separated by a very short space; wings, four, with numerous transverse nervures;

OUTLINE EYES OF BAËTIS, MAGNIFIED (Pictet).
1, Male (eyes nearly united).   2, Female (eyes remote).

the anterior long and straight, having complete nervation, and the costal edge angular; forceps large and well arched; two setæ, without any rudiment of a third.

Eaton:—Oculi of male divided into two unequal parts; the upper segment, cylindrical or somewhat turbinate, is faceted solely on the terminal surface; the lower and much smaller segment, oval in form, is annexed to the under orbit of the former, and is faceted all over with facets of less diameter of the turbinate part; hinder oculi large, foremost much smaller; wings anterior or mesothoracic, large, ovate-oblong, gradually rounded off from the terminal to the inner margin; hind-wings small or absent; neuration incomplete; sub-costa somewhat curved; median caudal seta aborted.

The genus *Cloëon* was separated from the *Ephemera* of Linnæus by Leach as having only two wings instead of four.

THE INSECTS.

Stephen defines *Cloëon*:—Head small, transverse; eyes *moderate*, remote lateral; thorax ovate; wings, *two*, elongate, rounded at the apex with numerous longitudinal nervures, and a few transverse ones; posterior pair wanting; abdomen rather long, attenuated at apex, with two more or less elongated setæ; legs slender, anterior pair frequently elongated; femora occasionally thickened; claws unequal.

Pictet:—Eyes of the male each surmounted by a great reticulated eye in the form of a turban—that is to say, borne on an abconical ring, at the base of which is the ordinary eye; wings, two or four, the superior (anterior) having very few transverse nervures, which ordinarily form

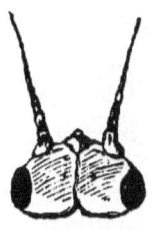 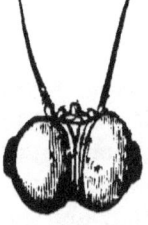

Female.   Male.
OUTLINE EYES OF CLOËON, MAGNIFIED (Pictet).

two bent lines. The inferior (posterior) wings, when they exist, are rudimentary, with nervation more or less incomplete; forceps of the male tolerably large; two caudal setæ, with a slight trace of the middle or third.

Eaton:—Hind-wings absent, intercalar rudimentary veinlets of the terminal margin single in a large majority of the species; forceps four-jointed, basal joint short, relatively stout; intermediate abdominal segments of female sub-equal in length; caudal setæ in male imago about twice, in female one to one and a half, as long as the body. In the sub-imago two-thirds and three-fourths as long.

Eaton has placed a genus, *Centroptilum*, between *Baëtis* and *Cloëon*, having four wings, the hind-wing oblique, elongate, and narrow, with the apex commonly obtuse, rarely acute, and usually with the costal projection acuminate; neuration limited to two simple longitudinal nervures; fore-wings devoid of colour ornamentation, free from cross veinlets in the marginal and sub-marginal areas; neuration as a whole similar to that of species of *Cloëon* or *Baëtis*. There are only two British species.

Centroptilum luteolum.—Wings very faintly grey tinted, sometimes very slightly tinted with the palest yellow ochre; setæ greyish-white or cinereous; turbinate eyes in male imago, bright light-red; simple eyes in female, greenish-black or black. This insect is the same as *Cloëon ochraceou* (Stephens), *Cloëon translucida* (Pictet). Eaton says it is abundant in Great Britain, and reaches maturity in our southern counties from April to November. No doubt one of the Olive Duns, *Centroptilum pennulatum*, has in the sub-imago state the wings a very little greyer than *C. luteolum*; distinguishable by its greater size. The male imago has the turbinate eyes of light cadmium orange, lower eyes olive-grey or black; those of the female imago olive-grey or greenish-black. According to Eaton, it is common in trout-streams from August to October.

Eaton says: "Among miscellaneous representatives of the genera, adult, and in good condition, *Cloëon* is easily distinguished by the absence of hind-wings; *Baëtis* by hind-wings broad and obtuse, with scarcely a cross veinlet at all; *Centroptilum* by the extreme narrowness of its very small wings, and usually by the slenderness of their costal projection. But to discriminate from *Cloëon* defective specimens of other genera deprived accidentally of their hind-wings is a task attended with insurmountable difficulty."

In *Cloëon* the turbinated large eyes, with scarcely any space between them, appear to be one of the chief characteristics of this genus. The insects themselves are also, as a rule, of very delicate structure, and of pale-yellow or ochreous tint.

The rudiment of a third seta can often be observed on examination with a high-power lens.

Stephens says that *Baëtis* can be distinguished from *Cloëon*, in the shape of the wings—those of *Baëtis* being long and narrow and somewhat pointed; those of *Cloëon* rounded at the apex.

In the genus *Cænis*, Stephens says: "This genus is readily known by the brevity of its wings and abdomen, the latter not exceeding the length of the thorax; its apex furnished with *three setæ*, by which character alone it may be distinguished from the two following genera—*Baëtis* and *Cloëon*."

According to Pictet, the head is short and large; eyes simple in both sexes, and always separated by a considerable space; wings, two, with very few transverse nervures; body short and thick; three setæ; very long in the male, very short in the female. These insects are very local, but occur occasionally in great swarms.

1. BAËTIS, WITH FORE WINGS ELONGATED AT APEX.
2. CLOËON, WITH FORE WINGS ROUNDED AT APEX.

The genus *Heptagenia* (Eaton).—The *Heptagenia sulphurea* is the *Baëtis elegans* of Stephens and Curtis, and *Baëtis cyanops* of Pictet, one of the Yellow Duns. *Heptagenia flavipennis.*—*Baëtis longicauda* (Stephens), *Ceræa* (Pictet), another light Yellow Dun, the "yellow uprights" of Devonshire.

The exact identification of the names by which artificial flies are known to anglers with the natural insects, particularly in the smaller genera and species of the *Ephemeridæ*, is no easy task.

The state and age of the insect, the time of year, the different atmospheric conditions, produce various shades of colour both in wings and body. At one time an olive tint, at another a yellow, or grey, or brownish-red.

The different nomenclature of the northern and southern counties is also another difficulty. In the north most of our southern duns are classed as drakes, the duns being taken chiefly from the *Phryganidæ*; whilst in the south the duns are almost all from the *Ephemeridæ*.

With few exceptions, we must, until each species has been more clearly defined and verified by microscopical investigations, rest satisfied to place the various duns, olives, quill-gnats, &c., in the genera, *Baëtis, Cloëon*, and *Potomanthus*.

It should be borne in mind that the different species of insects of the family *Ephemeridæ* we see as they emerge on the surface of, or float down, the water are for the most part in their sub-imago state. In their perfect or imago state they are hovering higher in the air (in the shape of various spinners) for the purpose of continuing their species; the female descending again from time to time to deposit her eggs; the male only occasionally, in his perfect state, being found on the surface.

The Turkey Brown of Ronalds, one of this family, with *three setæ*, is, we believe, the sub-imago of *Ephemera submarginata* (?) of Stephens; *Potomanthus geerii* (?) of Pictet; *Leptophlebia submarginata* of Eaton.[1]

The July Dun of Ronalds, another of this family, with *three setæ*, is the sub-imago of *Ephemera marginata* (Stephens); *Potomanthus marginatus* of Pictet.

The Hare's Ear is the sub-imago of *Baëtis bioculata*; the perfect insect being a spinner somewhat similar to the Jenny Spinner, *B. nyen*.

The March Brown or Dun Drake is the sub-imago of *Baëtis fluminum* (Pictet), or *Baëtis venosa* (Pictet and Stephens). The male imago is one of the early Red Spinners. In the "Scientific Angler" (p. 123, third edition) it is stated that "in May the March Brown is recognised as the Turkey Brown, light and dark, in accordance with the weather;" but the Turkey Brown has *three setæ*, the March Brown *only two*.

The various so-called Olive Duns may be classed chiefly in the genus *Baëtis*. Some have the wings much reticulated, in others the cross reticulations are very faint; some also have the turbinated eyes separated, others with them divided by a distinct line. There are a great many shades of Olives. *Baëtis lateralis*, the Olive Dun or Cocktail, is a good example of the reticulated wings.

The Pale Evening Dun of Ronalds, so prevalent in the

[1] Mr. Eaton, in Part V. of his monograph, published since, gives *Ecdyurus insignis* as the Turkey Brown of Ronalds.

autumn months, is the sub-imago of *Cloëon hyalanilum* or

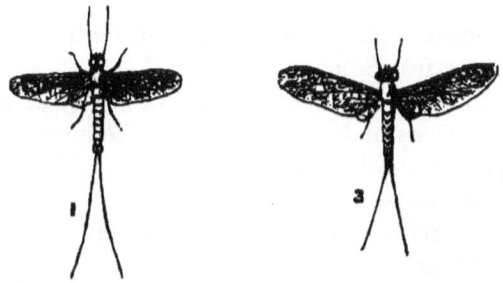

BAËTIS FLUMINUM.
1, Sub-Imago. 3, Female Sub-Imago.

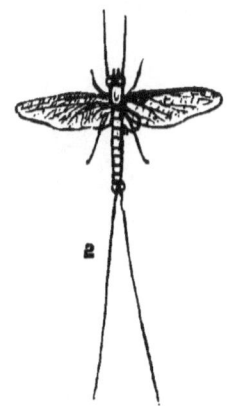

BAËTIS FLUMINUM.
2. Imago.

Ochraceum (Stephens); *Centroptilum luteola* (Eaton); the imago being a pale-coloured Spinner with a dark throat.

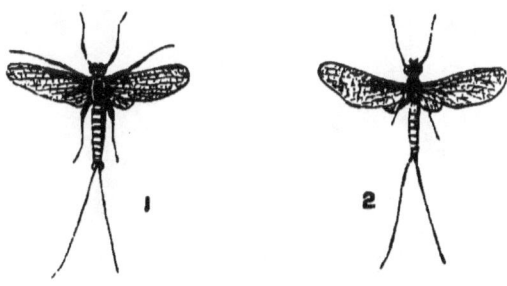

BAËTIS LATERALIS.
1, Male. 2, Female Sub-Imago.

The Quill Gnats (grey and red), which are not gnats at

all, are probably the sub-imago and imago state of *Cloëon dimidiatum*.

The August Dun, the sub-imago of *Baëtis autumnalis;* the imago a claret Spinner.

The Iron Blue is the sub-imago of the male of *Baëtis niger* of Linnæus and other authors; the imago being the Jenny Spinner, the female imago having an orange body, with light-hyaline wings.

The little Yellow May Dun of Ronalds may be *Baëtis longicauda* (Stephens).

CHAPTER XXI.

The order Hymenoptera (membrane-winged), metamorphosis complete, includes the Ants, Wasps, Bees, Saw-Flies, &c.

Of the Ants, the Large Red Ant (*Formica rufa*) and the Jet Ant (*Formica fuliginosa*) are the two species generally imitated by the fly-fisher.

In August and September, when, in vast numbers, the winged ants leave the ant-hills for the purpose of perpetuating their race, many—the males especially—fall on the surface of the water and become food for fish.

THE RED ANT.

Messrs. Kirby and Spence, in their introduction to "Entomology," give a most interesting account of the process by which the female ant gets rid of her wings. They say: "There is one circumstance occurring at this period of their history which affords a very affecting example of self-denial and self-devotion of these admirable creatures. If you have paid any attention to what is going forward in an ant-hill, you will have observed some larger than the rest, which at first sight appear, as well as the workers, to have no wings, but which, on closer examination, exhibit a small portion of their base or socket in which they are inserted. These are females that have cast their wings, not accidentally, but by a *voluntary* act. When an ant of this sex first emerges from the pupa she is adorned with two pair of wings, the upper or outer pair being larger than her body. With these, when a virgin, she is enabled to traverse the fields of ether surrounded by myriads of the other sex who are candidates for her favour; but when once the connubial rites are celebrated the unhappy husband falls and dies, and the widowed bride seeks only how she may provide for their mutual offspring. Panting no more to join the choir of aerial dancers, her only thought is to construct a subterranean abode in which she may deposit and attend to her eggs and cherish her embryo young. Her ample wings, which before were her chief ornament and the instrument of her pleasure, are now an encumbrance which incommode her in the fulfilment of the great duty uppermost in her mind; she therefore, without a moment's hesitation, plucks them from her shoulders. Might we not, then, address females who have families in words like those of Solomon, 'Go to the ant, ye *mothers*; consider her ways, and be wise'?"

It is generally in the warm days of August and September that the swarms of ants, both male and female, quit the nest and ascend into the air, the males rising first. The females are surrounded by hundreds of the males, and are wafted here and there by the slightest breeze. They sometimes rise to a very considerable height. It is after this *danse de l'amour* that, from various causes, either from cold or heavy showers, or from wind, or in the case of the males from pending death, these insects fall sometimes in countless numbers on the water, and are eagerly eaten by the fish. One author observes: "In the beginning of August I was going up the Orford river in Suffolk, in the evening, when my attention was caught by an infinite number of winged ants, both males and females, at which the fish were everywhere darting, floating alive on the surface of the water."

There are six species of British ants—the LARGE RED or HORSE ANT, *Formica rufa*; the JET ANT, *F. fuliginosa*; the RED ANT, *Myrmica rubra*; the COMMON YELLOW ANT, *F. flava*; the SMALL BLACK ANT, *F. fusca*; and the SLAVE-MAKING ANT, *F. sanguinea*.

The Saw-Flies have been divided into very many genera and species. They derive their name from the ovipositor of the female, which is often differently constructed in different genera. Some of these flies perforate the stalks of plants, and there deposit their eggs, the hole made by the ovipositor being filled with a frothy liquid, and sometimes a swelling like a gall-nut is formed.

THE INSECTS.

Newman says: "The common saw-flies, or *Atlantites*, have the antennæ forming a point. The club-horned saw-flies, or *Tenthredinites*, have the apex of the antennæ club-shaped."

The SAW-FLY of Ronalds is probably *Dolerus fulviventris* —a very common species, found everywhere, particularly on the plants known as horse-tails. It varies much in colour.

THE SAW-FLY.

THE GOVERNOR.

Another Hymenopterous insect—one of the bee family, *Andrena cingulata*—is, we think, the natural representative of that well-known lure for trout, the GOVERNOR. This insect is very plentiful in the spring and summer months, forming burrows in the sandbanks, and found constantly on the catkins of the sallow, and on other plants, from which it gathers quantities of pollen, and this, combined with its reddish abdomen, covered with small hairs, is an attractive morsel, should the insect, which often happens, be blown on the surface of the water.

In the order HEMIPTERA (half-winged) the membranous wings are covered by wing-covers, either entirely membranous and reflexed, or partly coriaceous and partly membranous and horizontal; the tarsi (terminal part of the leg) never more than three-jointed.

This order, comprising insects of very varied structure, is divided into two great divisions, *Heteroptera* and *Homoptera*, which by some authors are made into two distinct orders. The points in which they agree consist in the imperfect metamorphosis and the structure of the mouth.

The WATER-CRICKET (*Velia currens*) has a truncated thorax, with a reddish spot near the anterior margin; abdomen orange and black in upper part, orange below;

two rows of grey spots down the back. Ronalds says it runs upon the water and feeds upon small flies, and is one of the first insects the trout find there.

These insects do not swim, and the silky down with which their bodies are covered protects them from the water.

The WATER-SCORPION (*Nepa cinerea*), although generally found in stagnant waters, frequents also our slow-running rivers and mill-heads where mud accumulates. Newman says that the Nepites crawl on aquatic plants, but do not swim. They are all carnivorous and aquatic. The name is derived from the similarity of the fore-legs to the nippers of the true scorpion—one of the *Arachnidæ*. As this insect often uses its wings, it probably may drop occasionally on the water. The bright colour of the body and wings would probably attract a hungry trout.

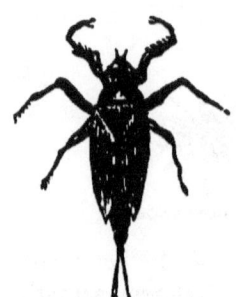

THE WATER-SCORPION.

The FROG-HOPPER (*Aphrophora spumaria;* sub-order, *Homoptera;* family *Circopidæ*), known in angling parlance as the Wren-Tail, is often a good lure for trout. There are two or three species, but the above is most common. Colour fuscous; antennæ placed between the eyes; hind-legs very long, enabling it to take long jumps. It is usually very prevalent in the hot days of June and July.

THE FROG-HOPPER (*A. bifasciata*).
1. The Insect.   2. The frothy excretion (cuckoo-spit).

One of the species (*A. bifasciata*) is common in our gardens, and its larvæ make that frothy excretion often seen on plants, and known as cuckoo-spit.

THE INSECTS.

The order DIPTERA (two wings), metamorphosis complete, have only one pair of membranous wings fixed to the centre of the sides of the thorax, but they also have two slender organs behind the wing, which, in entomological language, are called halteres, or balances. The mouth is provided with a sucker composed of six scaly lancet-like pieces. Body in three principal divisions, head, thorax, and abdomen. Legs, six. The eyes in the males are generally very large, uniting in the crown of the head, and in some occupy the entire head. This order comprises all the true flies—gnats, gad-flies, wasp-flies, ant-lions, &c.

Westwood says: "The immense profusion with which they are dispersed over the whole globe causes them to fulfil two very important functions in the economy of nature. First, they serve for food to a vast number of higher animals—swallows, fly-catchers, and many other birds; whilst, in the second place, they cease not in effecting the disappearance of all substances in a state of decomposition, both animal and vegetable." They are universal scavengers, and so great is their activity and the rapid succession of their generation, that Linnæus declared that these flies can consume a horse as fast as a lion. That pest to the fly-fisher, the grouse-shooter, and all who come in contact with it—the midge *Chironomus*—is of this order; and well may Westwood say: "Accordingly many species cease not to make man their prey."

The GOLDEN DUN MIDGE of Ronalds, who says it is a good fly on hot days, is *Chironomus plumosus*, one of the *Tipulidæ*.

The COW-DUNG FLY (*Scartophaga stercoraria*) is a well-known fly, and when blown on to the water has a very peculiar way of buzzing along the surface. The body is thickly covered with a yellow tomentum; head luteous; the eyes and proboscis are black; antennæ black; wings are greyish, with a slight tawny tinge along the borders of the veins; abdomen and legs tawny. These flies are plentiful in March and April, and at times the trout will take them freely.

The GRAVEL-BED or SPIDER-FLY is one of the family *Tipulidæ*. There are two species:—

1. *Anisomera nigra* (Ronalds says *obscura*). Body dark dunnish-black; wings grey; legs dull and ferruginous. Inhabits banks of the streams amongst the hills, and occurs most frequently in the north of England and in Scotland.

2. *A. villata.* Cinereous dull black; wings greyish; veins black and equally clouded; abdomen pilose on each side. Not so rare as *A. nigra,* and inhabits same localities (Walker).

The BLACK GNAT of Ronalds is the *Ramphamyia nigripes* of the family *Empidæ*.

The FISHERMAN's CURSE, the sight of which produces so many expressions of disgust and vexation, is of the same family, of which there are many species. The body is clothed with black hairs; wings colourless; legs blackish-brown in the male; the female is more hairy and darker, with legs varying in size.

The BLUEBOTTLE (*Musca vomitoria*) is sometimes used as a bait for trout, but it is more suitable for chub.

The COMMON HOUSE-FLY (*Musca domestica*) is at times a killing bait.

The DOWN-HILL FLY of Ronalds is of the family *Rhagionidæ*—*Rhagio scolopacis*.

The order LEPIDOPTERA includes the Butterflies and Moths. The various butterflies which cross our path when by the river-side rarely become food for fish—we have seen one occasionally taken by a voracious trout leaping quite out of the water after its prey—but our attention is often attracted and our eyes enchanted as we gaze upon some of these beautiful objects when they are culling the honey from the blossoms of the great meadow-thistle, or the sweet-scented dropwort, or the drooping honeysuckle. Spring-time, amongst many others, gives us the Sulphur or Brimstone; the Large White, with its black-tipped wing; the lovely Orange-Tip, with its rapid flight; the Wood-Lady; the Gate-Keeper or Speckled-Wall. The summer months of June, July, and August are enlivened by the

THE INSECTS. 353

beautiful forms of the Red Admiral, the Large Tortoise-shell, the Painted Lady, the lovely Fritillarias, the magnificent Peacock, and many others. All these at times and in their seasons will be found, without much searching, amongst the many objects we can see if we only look for them.

Many of the moths, both in their larva and imago state, are taken by fish when, by accident or otherwise, they drop on the surface of the water.

The larva of the Tiger-Moth (*Arctia caga*) is the RED PALMER of Ronalds—the WOOLLY BEAR of other authors.

The larva of the Spotted Ermine Moth (*Spilosoma lubricepeda*) is the BROWN PALMER. There is, however, another brown caterpillar, the larva of the *Cycnia mendica*, which feeds on aquatic plants, which also is taken by trout.

The larva of the Fox-Moth (*Laciœampa rubi*) is the BLACK PALMER. It is generally found on brambles, and is known in Hampshire by the name of the Devil's Gold Ring, and is the Scotsmen's "Hairy Oubit."

The white moths, which are found at times near rivers, are of three kinds:—

1. The SATIN MOTH (*Stilpnotia salicis*), with white body.
2. The WATER ERMINE MOTH (*Spilosoma papyratia*), with yellow body.
3. *Potheria auriflua*, with dark-brown purplish body, which is probably the COACHMAN of the fly-fisher.

The male of the DRINKER-MOTH (*Odonestus potatoria*), of a dark reddish-buff, clouded with fulvous, would, if well imitated, make a good night-fly.

The ALDER-MOTH (*Acronycta alni*) makes a famous night-fly. Wings greyish-brown, inner margin broadly suffused with black; head-wings whitish-buff; legs grey. An imitation of this fly was made by Jones of Jermyn Street many years since, and went by the name of the CARSHALTON DESTRUCTOR. We have tried it with marked success on all rivers as a late evening-fly.

A good brown fly for night might be made to imitate the COMMON TISSUE MOTH (*Triphora dubitata*).

A very beautiful moth, the BULL-RUSH MOTH (*Monagria typhæ*), may be often seen flying round and settling on what is usually called the Bull-Rush, the proper name being the Reed-Mace or Cat's Tail (*Typha latifolia*). It is of a beautiful pale buff-colour, and often on a fine still evening a great many of them collect around and on the stems and flowering parts.

The PLUME-MOTHS, particularly that large white, floating, plumy *Pterophorus pentadactylus*, are often observed in the hot evenings of June and July.

Before leaving the subject of the Insects, we would draw attention to that peculiar white filmy substance known as *gossamer*, which often floats by us on a fine August day, sometimes in such quantities as to cover our clothes with its silvery threads. Those who are abroad in the early morning will, at times, find the meadows interlaced with innumerable thin lines crossing and recrossing each other, covered with dewdrops sparkling like diamonds in the sunshine. These gossamer threads are produced by small spiders, not always of the same species, but why so produced has never been quite satisfactorily answered. Some suppose they are formed for the collection of the dew, which the little animals eagerly drink; others, for the purpose of wafting themselves into the upper air, to seek for the small insects to be found there. The single threads are often so fine as to be almost imperceptible. The spiders which produce them shoot them out from their spinnerets, and being caught by the ascending current of heated air, are borne up, taking the spider with them. That careful observer of nature, Gilbert White, in a letter to Daines Barrington, says: "Every day in fine weather, in autumn chiefly, do I see these spiders shooting out their webs and mounting aloft. They will go off from your finger, if you will take them into your hand. Last summer one alighted on my book as I was reading in the parlour, and running to the top of the page and shooting out a web, took its departure from thence. But what I most wondered at was, that it went off with considerable velocity in a place where no air was stirring, and I am sure I did not assist it with my breath; so that these little creatures seem to have, while mounting, some locomotive power without the use of wings, and to move in the air faster than the air itself."

The formation of these gossamer webs is, indeed, a wonderful provision to enable these tiny aeronauts—these little creatures without wings—to ascend into the upper air for some specific purpose; this end accomplished, they are enabled by coiling these webs round their bodies again to descend to earth.

In early times gossamer was supposed to be " scorched dew." Spencer says :—

> "The fine nets which oft we woven see
> Of scorched dew."

Henry More, however, who wrote about 1640, suspected that spiders were the cause :—

> "As light and thin as cobwebs that do fly
> In the blew air, caus'd by the autumnal sun
> That boils the dew that on the earth doth lie,
>   May seem this whitish rag then is the scum ;
> Unless that wiser men make 't the field spider's loom."

CHAPTER XXII.

OF THE EARTH-WORMS (ANNELIDÆ).

A REVIEWER of Darwin's book, "Vegetable Moulds and Earth-Worms," remarks: "In the eyes of most men the earth-worm is a mere blind, dumb, senseless, and unpleasantly slimy annelid. Mr. Darwin undertakes to rehabilitate his character, and the earth-worm steps forth at once as an intelligent and beneficent personage, a worker of vast geological changes, a planer-down of mountain-sides, a friend of man, and an ally for the Society for the Preservation of Ancient Monuments" ("Life," vol. iii. p. 213). The book itself ("The Formation of Vegetable Mould through the Action of Worms," 1 vol. octavo) is full of most interesting facts on their habits, food, digestive structure, intelligence, manner of prehension of objects and protection of their burrows, amount of earth brought up by them, &c.

A short notice of their structure and natural history may be useful to many who are, perhaps, quite unaware that these creatures are so perfectly organised:—

"The body of a large earth-worm consists of from one hundred to two hundred almost cylindrical rings or segments, each furnished with minute bristles. The muscular system is well developed. Worms can crawl backwards and forwards, and by the aid of their affixed tails can retreat with extraordinary rapidity into their burrows. The mouth is situated at the anterior end of the body, and is provided with a little projection (lobe or lip), which is used for prehension; and behind the mouth is a strong pharynx which leads to the œsophagus, on each side of which, in the lower part, are three pairs of glands which secrete a surprising amount of carbonate of lime; nothing like these glands is known in any other animal. The œsophagus, lower down, is enlarged into a crop in front of the gizzard, which opens into an intestine of a very remarkable structure."

Worms breathe by the skin, as they do not possess any special respiratory organs. The two sexes are united in the same individual, but two individuals pair together, mutual fecundation taking place by means of the thickened knot (*clitellum*) on the body. The nervous system is fairly well developed, and the two almost confluent cerebral ganglia are situated very near the anterior end of the body.

Worms are destitute of eyes; and Hoffmeister states that they are, with the exception of a few individuals, extremely sensitive to light. Darwin did not always find this to be the case.

Worms do not possess any sense of hearing, but are extremely sensitive to vibrations in any solid object.

"The whole body of a worm is sensitive to contact. A slight puff of air from the mouth causes an instant retreat."

"Worms are very timid. It may be doubted whether they suffer as much pain when injured as they seem to express in their contortions. Judging by their eagerness for certain kinds of foods, they must enjoy the pleasure of eating. Worms are omnivorous. They swallow an enormous quantity of earth, out of which they extract any digestible matter which it may contain. They also consume a large number of half-decayed leaves of all kinds."

Of the eleven species of British earth-worms, those principally used by anglers are :—

1. *Lumbricus terrestris*, the EARTH-WORM, LOB, or DEW-WORM of Hofland, Stoddart, and Younger. The skin reflects a beautiful iridescence, more especially from the dorsal margins of the segments; the basis of the minute spines placed on each side of the body is of a clear brown. It is common everywhere, and delights in a wet, loose soil.

2. *Lumbricus minor*, the MARSH-WORM or RED-HEAD of Stoddart, RED-WORM of Hofland, peacock-red or black-headed RED-WORM, SEGG-WORM, the TROUT-WORM. The skin is not iridescent, or only very slightly so. Found in

wet gravelly ground, on the sides of rivers and burns, under the masses of confervæ, &c., on the front of rocks over which water trickles. Very common; a favourite bait for trout.

3. *Lumbricus anatomicus*, BLACK-HEAD or BUTTON-WORM (Stoddart); the black-headed small TAIL-WORM (Younger). The anterior portion of the body of a uniform dull umber-brown; posterior portion pale orange-brown. Very common in meadows and gardens, and much used for bait.

4. *Lumbricus fœtidus*, the BRANDLING (Hofland, Stoddart), BRANDLING or BRAMBLE-WORM (Younger). Body banded with alternate brown and yellow segments. Two abbreviated impressed lines on the second segment behind the head. It exhales a disagreeable odour, of which it is difficult to rid oneself. Found in very old dung-heaps. Much esteemed by the angler.

5. The *Lumbricus viridis*, with greenish body; dull and inactive; throwing itself into an imperfect coil when disturbed. Found under stones in pastures and by the burnside, and under the dried droppings of cattle; is generally rejected by the angler as bait for fish.

Some of the family *Hirudinidæ* (Leeches) may be observed at times. Although the medicinal Leech (*Hirudo medicinalis*) is a British species, it is very rare, and not likely to come under our notice. Not so with the HORSE-LEECH (*Hæmopsis sanguisuga*), which is frequently found in ponds and lakes. It is much larger than the medicinal species, but its teeth are comparatively blunt; it feeds greedily on earth-worms. It is of a greenish-black colour on the dorsal, and yellowish green on the ventral, surface, marked with irregular spots; eyes indistinct.

A number of small leeches of the family *Glossoporidæ* (genus *Glossiphinia*), many of them transparent, so that the viscera may be seen, inhabit our ponds and streams. They are tenants of pure fresh-waters, incapable of swimming. One is the chequered leech (*G. tessellata*), found in weedy ponds; another (*G. sexoculata*) is found in lakes, ponds, and rivulets; common in the river White Adder, Berwickshire.

CHAPTER XXIII.

RIVER-SIDE FLOWERS.

> " How often doth a wild-flower bring
> Fancies and thoughts that seem to spring
> From inmost depth of feeling!
> Nay, often they have power to bless
> With their uncultured loveliness,
> And far into the aching breast
> There goes a heavenly thought of rest,
> With their soft influence stealing."

IN describing the various plants, we have endeavoured to avoid a too scientific definition, giving only the general character, with a reference to the number of the coloured illustrations in Sowerby's "English Botany," third edition. The general characters are taken chiefly from Hooker's "British Flora." As far as possible, they are placed in the order of the months in which they flower, not in any botanical arrangement. Hooker says: "Be assured that in plants taken individually and in an isolated manner there are subjects that will give ample scope for the employment of the talents of the greatest philosophers, in the due contemplation of which they may derive both pleasure and advantage to themselves and be the means of communicating them to others."

> "The well-directed sight
> Brings in each flower an universe of light."

Independent of the water-side plants, there are others which meet the eye on our way, both in meadow and copse, such as the primrose, violet, blue-bell, daffodil, buttercup, daisy, and snowdrop:—

"Buttercups and daisies, oh! the pretty flowers,
Coming ere the spring-time, to tell of sunny hours;
While the trees are leafless, while the fields are bare,
Buttercups and daisies spring up here and there."

The BUTTERCUP or UPRIGHT MEADOW CROWFOOT (*Ranunculus acris*)—Fig. in "English Botany," 33—with its bright golden-yellow shining flowers, is very common. The whole plant is very acrid. Curtis says that even pulling up the plant will at times blister the hands. Sheep and goats will eat it, but cows and horses refuse it. The leaves pounded and used as a poultice will produce vesication. Topically applied, it is supposed to be efficacious in rheumatism, and has been employed as a blister.

The DAISY (*Bellis perennis*)—Fig. in "E. B.," 722. The word daisy is a compound of *day's* and *eye*—day's eye. In Yorkshire it is called Bairnwort, probably from the delight which children take in gathering these flowers. The French call it *Marguerite*, expressive of beauty, from *margarita*, a pearl. "The daisy," says Mr. Phillips, "has been made the emblem of innocence, because it contributes more than any other flower to infantile amusement and the joys of childhood. In the days of chivalry it was the emblem of fidelity in love, and was frequently borne at tournaments both by ladies and by knights." This little modest crimson-tipped flower appears ever to have been a general favourite. "Who," says Miss Kent, "can see or hear the name of daisy, the common field daisy, without a thousand pleasurable associations! It is connected with the sports of childhood and with the pleasures of youth. We walk abroad to seek it, yet it is the emblem of home. It is a favourite with man, woman, and child; it is the *robin* of flowers. Turn it all ways and on every side you will find new beauty. You are attracted by the snowy-white leaves (florets of the ray), contrasted by the golden tuft in the centre (florets of the disk), as it rears its head above the green grass; pluck it and you will find it backed by a delicate star of green (involucrum) and tipped with a blush-colour of bright crimson.

"'Daisies with their pinky lashes'

are among the first darlings of spring. They are in flower

almost all the year; closing in the evening and wet weather, and opening on the return of the sun."

Leyden says :—

> "Star of the mead! sweet daughter of the day,
> Whose opening flower invites the morning ray,
> Oft have I watched thy closing buds at eve,
> Which for the parting sunbeams seem to grieve,
> And, when gay morning gilt the dew-bright plain,
> Seen them unclasp their folded leaves again."

The "PALE PRIMROSE, Harbinger of Spring" (*Primula vulgaris*)—Fig. in "E. B.," 1129—which decks the banks and copses so gloriously with its petals of delicate yellow, often so abundant as completely to carpet the ground on which it is growing, and frequently combined with the trailing periwinkle and the purple violet. Wordsworth says :—

> "Through primrose tufts in that sweet bower,
> The periwinkle trail'd its wreaths,
> And 'tis my faith that every flower
> Enjoys the air it breathes."

If we trace the flower-stalks of the primrose to their base, we shall find that they all arise from one common point, and constitute, as it were, a sessile umbel. The wanton destruction of the primrose for what is now called Primrose Day is appalling, and will, if not checked, annihilate this lovely emblem of the spring.

Another primula is the COWSLIP or PAIGLE (*Primula veris*)—Fig. in "E. B.," 1730—which frequents the spring-meadows more than the copses and hedgerows, and differs from the primrose in having a regular umbellated flower-stalk. Whether the cowslip is a distinct species or merely a variety of the common primrose, or *vice versâ*, is a question. The late Professor Henslow has seen both produced from the same root, but it is to be noticed that the varieties or species are very constant in their wild state, both as to their forms and localities. The five orange spots at the bottom of the corolla are supposed to contain the peculiar odour of the plant.

Shakespeare alludes to this in "A Midsummer Night's Dream" (Act ii. sc. 1):—

> "The cowslips tall her pensioners be:
> In their gold coats spots you see;
> Those be rubies, fairy favours,
> In those freckles live their savours."

Milton in "Lycidas" notices the drooping flowers of the cowslip:—

> "With cowslips wan, that hang the pensive head."

Montgomery is particularly happy in his allusion to this plant:—

> "Now in my walk with sweet surprise
> I see the first spring cowslip rise,
>   The plant whose pensile flowers
> Bend to the earth their beauteous eyes
>   In sunshine as in showers."

Cowslip tea and cowslip wine were formerly much used by country-folks. The latter has a peculiar musk-like flavour, and is supposed to favour sleep:—

> "Where thick thy primrose blossoms play,
>   Lively and innocent as they,
>   O'er coppice, lawns, and dells,
>   In bands the village children stray
>   To pluck thy honied bells,
> Whose simple sweets, with curious skill,
> The frugal cottage dames distil,
>   Nor envy France the vine,
> While many a festal cup they fill
>   Of Britain's homely wine."

Mixed with the primrose in coppice or on the banks, the pale-blue scentless VIOLET (*Viola hirta*)—Fig. in "E. B.," 172—or HAIRY VIOLET, is often found, particularly on chalky soils. Although closely allied, it is distinguished from the SWEET VIOLET (*Viola odorata*) by its pale colour; by the short, not creeping shoots; by the greater hairiness of the plant; and by the situation of the little bracteas of the scape (stalk), which are below, while in *V. odorata* they are above the middle. The sweet violet is occasionally met with; its flowers are more of a deep purple-blue, often white. According to Hasselquist, the *sorbet* of the Turks is prepared from these flowers mixed with sugar. The sweet violet is very rare in Scotland, and the Highland ladies of former days made a cosmetic of it. There is an old Gaelic saying, "Anoint thy face with goat's milk in which violets have been infused, and there is not a young prince upon earth who will not be charmed with thy beauty."

The DOG VIOLET (*V. canina*) is distinguished from having a flower-stem.

The WOOD SORRELL (*Oxalis acetosella*)—Fig. in "E. B.," 310—with its handsome white drooping flowers, is another beautiful little plant, found in the coppices in the spring.

The flowers of the WILD HYACINTH or BLUE-BELL (*Hyacinthus nonscriptus*)—Fig. in "E. B.," 1523—purple many of our woods about the end of April and beginning of May. This flower is dedicated to the patron saint of England, St. George.

The COMMON DAFFODIL (*Narcissus pseudo-narcissus*)—Fig. in "E. B.," 157—is another of the yellow flowers of spring; it grows in great profusion in moist woods, meadows, and orchards. Flowers in March and April:—

"Daffodils,
That come before the swallow dares, and take
The winds of March with beauty."
—*Winter's Tale*, Act iv. sc. 3.

"When daffodils begin to peer
With heigh! the doxy over the dale,
Why, then comes in the sweet o' the year,
For the red blood reigns in the winter's pale."
—*Winter's Tale*, Act iv. sc. 1.

All these, with the "chaste snowdrop, venturous harbinger of spring," form a parterre delightful to the fisher's eye and difficult to surpass. Those who follow in the spring and summer months the streams of the north, "o'er mountain and moor," will often come across that curious insectivorous plant the VENUS'S CATCHFLY (*Drosera*); or the

Bog Asphodel (*Narthecium*), with its yellowish-orange flower-scape; or the lovely Grass of Parnassus (*Parnassus Palustris*)—

"Whose modest bloom sheds beauty o'er the lonely moors."

One of the greatest ornaments of the banks of streams and adjacent water-meadows in the spring is the Common Marsh Marigold (*Caltha palustris*)—Fig. in "E.B.," 40—from the greek word καλξθος, a cup. It is also known as Gowans, Golden-Knobs, Meadow-Bouts, Mare-Blobs.

The stems rise from twelve to eighteen inches, round, hollow, and smooth. Leaves large, somewhat heart-shaped; lower ones on long stalks, the upper attached close to the stem; flowers bright yellow. It generally flowers in March and April, but in forward springs as early as February, and where it grows in abundance, the water-meadows and margins of the streams glow with its brilliant yellow blossoms. In some counties the people strew the flowers before their doors and weave them into wreaths on May-day. The flower-buds preserved in salted vinegar have been used as substitutes for capers. A small variety is sometimes found in marshy places:—

> "And see the flaunting marigold,
> Gay from its marshy bank unfold
> 'Mid minor lights its disks, that shine
> Like suns of brightness."

In some stagnant waters the Spiked Water Milfoil (*Mynophyllum spicatum*)—Fig. in "E. B.," 514—rears its slender, much-branched stem. This must not be confounded with the Common Milfoil (*Achillea*), of a different class and order, and so also another monœcius plant, the Vernal Water Starwort (*Callitriche verna*), with the Sea Starwort or Michaelmas Daisy (*Aster trifolium*), a syngenecious plant.

The Common Butter-Bur (*Petasitis vulgaris*)—Fig. in "E. B.," 119—name derived from the Greek word πετασος, a covering for the head, or umbrella, in allusion to the large leaves. From the early flowering of the plant before the leaves appear, the rod-fishers, as a rule, only see the leaves (the flowers having died away), which are the largest of any British plant, and when full-grown are nearly three feet in diameter. They all spring from the root and stand on thick upright foot-stalks, rounded and heart-shaped, somewhat toothed, yellowish-green above, downy, but not very white underneath. The flowers are pinkish—more of a pale flesh-colour. This plant is sometimes known by the name of Pestilent Wort, from the juice being supposed to be able to cure the plague. The name Butter Bur is derived from the habit of the milk-maids always using the large leaves to wrap the butter in. Flowers April and May.

The GREAT SEDGE or CAREX (*Carex riparia*)—Fig. in "E. B.," 1679—is very common on the banks of most rivers, and when flowering, if touched by the line or rod, sends out a great cloud of pollen from its dark purplish-yellow spikes. The name is derived from the Greek word κειρο, to shear or cut, in allusion to its sharp leaves and stems. It is distinguished from the LESSER SEDGE (*Carex paludosa*)—Fig. in "E. B.," 1678—by the leaves being broader and more deeply keeled and rougher; also by the acuminated scales of its sterile spikes. Flowers May.

There are sixty-three British species of *Carex*.

A very pretty little plant, common on the banks of lakes and ornamental waters, and often found in our smaller streams, is the MARSH PENNY-WORT, WATER-CUP, WHITE-ROT (*Hydrocotyle vulgaris*)—Fig. in "E. B.," 566—from the Greek words *udor*, water, and *cotule*, a cup. The stems are creeping, thread-shaped, and tender and smooth, running on the ground, rooting at each joint, producing a tuft of leaves and flowers. The leaf-stalks rise from two to four inches; leaves smooth, glossy, light-green, somewhat cup-shaped, a little depressed, *with a white dot in the centre*. Flowers May, June. Baxter says: "The whole plant is acrid, and probably, like others of the umbelliferous tribe growing in wet places, poisonous.

This plant has received its English names of White-Rot, Flukewort, Sheep-Killing Penny-Grass, Sheep's Bane, and Penny-Rot from an old belief that feeding upon it causes the liver-rot in sheep; this opinion, which is altogether an error, arose from the fluke or flounder insect (*Fasciola hepatica*) being found in marshy grounds where this plant abounds, but it is well-known that sheep never eat this plant.

In the month of June a great alteration takes place in the colours of many of the flowering plants. The brilliant golden-yellows of the spring months have for the most part disappeared, or are intermingled with others equally as beautiful. Clare, in his "Wanderings in June," says:—

> "How strange a scene has come to pass
>   Since summer 'gan its reign,
> Spring flowers are buried in the grass,
>   To sleep till spring again!
>
> Her dewdrops evening still receives,
>   To gild the morning hours,
> But dewdrops fall on open'd leaves,
>   And moisten stranger flowers.
>
> The artless daisy's smiling face,
>   My wanderings find no more,
> The king-cups that supplied their place,
>   Their golden race is o'er."

June, with its May-flies and flowers, with its warm days and health-giving breezes, entices us away from the murky city, and to respond to the "Summer Call:"—

> "Come away! The summer hours
> Woo thee far to founts and bowers;
> O'er the very waters now,
>   In their play,
> Flowers are shedding beauty's glow—
>   Come away!
> Where the lily's tender gleam
> Quivers on the glancing stream,
>   Come away!
>
> All the air is filled with sound,
> Soft and sultry, and profound,
> Murmurs through the shadowy grass
>   Lightly stray.
> Faint winds whisper as they pass—
>   Come away!
> Where the bee's deep music swells
> From the trembling foxglove bells,
>   Come away!"

One of the most common of water-plants is the WATER CROWFOOT (*Ranunculus aquatilis*)—Fig. in " E. B.," 17. It is found in almost every river, lake, or pond. The flowers are white, with yellow centre (stamens). The leaves which grow under the water are two or three forked. The floating leaves are three-sided, with cut lobes. It flowers in May and June. There are several varieties of the Water Crowfoot. That which is mostly found in our fast-running streams is the *Ranunculus pseudo-fluitans*, and may be distinguished from the above in the submerged leaves being more like tassels, and in often having no floating leaves.

The WATER SWEET-GRASS or WHORL-GRASS (*Catabrosa aquatica*)—Fig. in " E. B.," 1750—from the Greek word *katabrosis*, a gnawing, from the gnawed-like appearance of the husks—is a very pretty, graceful grass—found on the margins of our ponds and rivers. It is one of the sweetest of our grasses, and all kinds of water-fowl are particularly fond of the young shoots. The stems when growing in the water are partly floating. The leaves are strap-shaped, nearly flat, bright green in colour, smooth except at their margins. The panicles (*i.e.*, assemblage of flowers growing on divided stalks, as in ' the oat) are alternate. Flowers May and June.

The COMFREY (*Symphytum officinale*—Fig. in " E. B.," 1115—from the Greek *sumphuo*, to grow together, from its supposed power of healing wounds—with its hanging clusters of purple and white flowers, and rough stems and leaves, is well known to many who frequent the river's bank. The stem rises to two or three feet in height; the leaves at the root are on long foot-stalks, and are very rough; the clusters of purple and white flowers grow in pairs and are drooping. The colours vary much. It is often found with yellowish-white flowers only. The root abounds in a pure mucilage, which renders it useful in coughs and all internal irritations. The leaves give a grateful flavour to cakes and panada. Cows and sheep eat it; horses, goats, and swine refuse it. A decoction of the roots is used by dyers to extract the colouring-matter from gum lac. Flowers May and June.

The BUR-MARIGOLD (Bidens). There are two species, met with chiefly on the margins of lakes and ponds and small rivulets. The *Bidens tripartita* (Fig. in "E. B.," 94), also known as the Three-Lobed Bur-Marigold, Bastard Agrimony, Water-Hemp, Double-Foot. The stem rises from one to three feet in height; leaves pointed, dark-green, and strongly toothed, divided into three parts. The flowers are solitary, terminal, brownish-yellow, and somewhat drooping. Flowers July.

The NODDING BUR-MARIGOLD (*Bidens cernua*)—Fig. in "E. B.," 93—has undivided leaves and larger flowers. Flowers June to August.

The CELERY-LEAVED CROWFOOT (*Ranunculus sceleratus*)—Fig. in "E. B.," 27—is found by the margins of pools and ditches. The stem is stout and succulent, leaves glabrous, lower ones broad and glossy. Flowers extremely small and pale yellow. The plant grows one to two feet high. Flowers June.

The GREAT WILD VALERIAN (*Valeriana officinalis*)—Fig. in "E. B.," 666—from *valeo*, to be powerful—is very abundant on the banks of our rivers and pools. The plant grows from three to four feet in height. Its lower leaves are on long foot-stalks. Its pale flesh-coloured flowers on the high green flowering-stalk render it very conspicuous.

The root is tuberous, very aromatic. Cats are very fond of the scent of the root, and it will also attract rats. When cats rub themselves on this root, is it for the purpose of attracting the rats? The leaves are much used by the poor as an application to fresh wounds, and it is sometimes called All-Heal. Flowers June and July.

The SWEET-FLAG or SWEET-RUSH (*Acorus calamus*)—Fig. in "E. B.," 1391—is not very common in the southern and western districts. It is found on the banks of rivers in the middle and south-eastern counties, and plentiful in Norfolk and Suffolk. We have found it on the Itchen, Test, Kennet, and Thames. The root is very aromatic, with a bitter acrid taste. The leaves upright and sword-shaped, bright green in colour, sheathing one another, about an inch broad, and from two to three feet high.

The sheathed fruit-stalk is tapering, and closely studded with numerous small green flowers set in spiral lines. The name *acorus* is derived from the Greek words, *a*, without, κοριον, the pupil of the eye, from its supposed power of

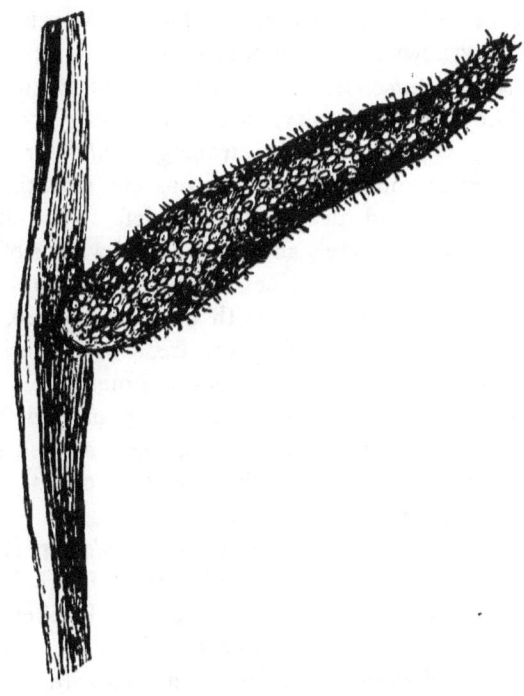

THE SWEET FLAG.
The flowering stem and flowers.

curing diseases of this organ. Its agreeable scent has recommended it for garlands, and on festival days it used to be strewed on the floor of the Cathedral at Norwich.

"The Mayor of Norwich holds in June
His annual feast and show,
And to the grand Cathedral Church
Processions with him go.

"And they, the grey and solemn aisles,
And all the ancient floor,
Are with the aromatic leaves,
Bestrewed thickly o'er.

The dried root powdered is used by the country people in Norfolk for the cure of the ague, and the candied fresh root is said to be employed at Constantinople as a preventive against epidemic diseases. It is also employed by perfumers in the making of hair-powder. Flowers in June.

A very beautiful and elegant plant—the WATER-VIOLET, WATER GILLYFLOWER, or MARSH HOTTONIA (Fig. in "E. B.," 1128), *Hottonia palustris,* named after Pierre Hotton, professor at Leyden University, is found in many pools and moist places. The stems are trailing, round, and leafy. The leaves grow under water, numerous and smooth, bright green in colour, and deeply feather-cleft. The flowery stems rise above the water, and bear several whorls of large, handsome, rose or pink coloured flowers (8–10 in each whorl), one above the other, with a yellow centre. The leaves are said to be eaten by the fresh-water periwinkle and other small shell-fish. Flowers in June.

Considerable interest has of late been excited by the discovery of new-born fish (first, we believe, noticed by Mr. G. E. Simms, of Oxford) in the bladders of the GREAT BLADDERWORT (*Utricularia vulgaris*).

There are three British species—*U. vulgaris* (Fig. in "E. B.," 1125), *U. intermedia* (Fig. in "E. B.," 1127), and *U. minor* (Fig. in "E. B.," 1126). Hooker says: "The British species of this genus are all aquatic, and their roots, stems, and even leaves, are furnished with numerous membranaceous reticulated vesicles, which, according to Hayne, are filled with water till it is necessary the plant should rise to the surface and expand its blossoms above that fluid. The vesicles are then found to contain only air, by aid of which the plant floats. This air again in autumn gives place to water, and the plant descends to ripen its seeds at the bottom." Mr. Wilson observes on the bladders of Great Bladderwort or Hooded Milfoil (*U. vulgaris*), that "they have an orifice closed by an elastic valve, opening inwards, and of much thinner texture than the bladder to which it is attached where the crest is placed. Aquatic insects often enter these bladders, and are, of course, confined there."

Mr. Darwin ("Insectivorous Plants," pp. 428 *et seq.*) says of *Utricularia vulgaris*, the common Bladderwort: "Living plants from Yorkshire were sent me by Dr. Hooker. Five bladders containing prey of some kind were examined. The first included five cypris, a large copepod, and a Diaptomous; the second four cypris; the third a single rather large crustacean; the fourth six crustaceans; and the fifth ten. My son examined the quadrifid processes in a bladder containing the remains of two crustaceans, and found some of them full of spherical or irregularly-shaped masses of matter, which were observed to move and to coalesce. These masses, therefore, consisted of protoplasm." Of *Utricularia minor*, the Lesser Bladderwort—a rare plant in England, but frequently found in Scotland—Darwin says: "This rare species was sent to me from Cheshire. The plants were collected in the middle of July, and the contents of five bladders, which from their opacity seemed full of prey, were examined. The first contained no less than twenty-four minute fresh-water crustaceans, most of them consisting of empty shells, or including only a few drops of red oily matter; the second contained twenty; the third fifteen; the fourth ten, some of them being rather larger than usual; and the fifth, which seemed stuffed quite full, containing only seven, but five of these were of an unusually large size. The prey, therefore, judging from these five bladders, consists exclusively of fresh-water crustaceans, most of which appeared to be distinct species from those found in the bladders of the former species."

Mr. Darwin also quotes from Mrs. Treat, of New Jersey, who had examined a number of bladders of a North American species (*Utricularia clandestina*), in which a vast number of captured animals were found within the bladders, some being crustaceans, but the greater number delicate, elongated larvæ, I suppose of Culicidæ. On some stems fully nine out of every ten bladders contained the larvæ or their remains; the larvæ showing signs of life from twenty-four to thirty-six hours after being imprisoned, and then perished.

At page 438 he says: "Finally, as numerous minute

animals are captured by this plant (*Utricularia neglecta*) in its native country, and when cultivated there can be no doubt that the bladders, though so small, are far from being in a rudimentary condition; on the contrary, they are highly efficient traps. Nor can there be any doubt that matter is absorbed from the decayed prey by the quadrifid and bifid processes, and that protoplasm is thus generated. What tempts animals of such diverse kinds to enter the cavity beneath the bowed antennæ, and then force their way through the little slit-like orifice between the valve and collar into the bladders filled with water, I cannot conjecture."

That the plant is nourished by the various insects and animals it is able to capture seems to be beyond doubt. Darwin, writing of the *Utricularia neglecta*, says: "The real use for the bladders is to capture small aquatic animals, and this they do on a large scale. In the first lot of plants I received from the New Forest early in July, a large proportion of the fully-grown bladders contained prey:" and he also states that a plant of *Utricularia vulgaris*, which had been in almost pure water, was placed by Cohn one evening into water swarming with crustaceans, and by the next morning most of the bladders contained those animals entrapped and swimming round in their prisons. Fresh-water worms were found by Cohn in some bladders. Darwin describes the mode in which these different forms enter the bladders. He says: "Animals enter the bladders by bending inwards the posterior free edge of the valve, which, from being highly elastic, shuts again instantly. As the edge is extremely thin and fits closely against the edge of the collar, both projecting into the bladder, it would evidently be very difficult for any animal to get out when once imprisoned, and apparently they never do escape. To show how closely the edge fits, I may mention that my son found a Daphne which had inserted one of its antennæ into the slit, and it was thus held fast during a whole day. On three or four occasions I have seen long narrow larvæ, both dead and alive, wedged between the corner of the valve and collar, with half their bodies within the bladder and half out" (Darwin, "Insectivorous Plants," pp. 405–406).

The genus *Utricularia* has a calyx of two-leaved equal. Corolla personate, spurred; capsule globose, of one cell; named from *utriculus*, a little bladder.

Utricularia vulgaris has the spur conical, upper lip as long as the projecting palate, leaves pinnato-multifid. Roots much branched, shooters or runners floating horizontally in the water, clothed with capillary multifid leaves, bristly at the margin, being little cristate bladders. Scape erect, four to six inches high; six or eight bright yellow flowers in a raceme. Flowers June and July.

Utricularia minor, Lesser Bladderwort, spur extremely short; upper lip as long as the palate, leaves subtripartite. Vesicles mixed with the leaves, which latter are glabrous at the margin. Flowers June and July.

Utricularia intermedia. Very similar to the *Vulgaris*. Flowers are smaller, of a paler yellow, and have a long upper lip; the flowers are more leafy, and the bladders arise from branched stalks, not from the leaves. It is more rare. Flowers June and July.

There are fourteen British species of the POND-WEEDS, *Potamogeton* (from the Greek words, *potomos*, a river, and *geiton*, a neighbour), all growing in the water. Those which are commonly found are the sharp-fruited and broad-leaved pond-weed, *P. natans;* the fennel-leaved pond-weed, *P. pectinalis* (Fig. in "E. B.," 1405); the grassy-leaved pond-weed, *P. gramineus* (Fig. in "E. B.," 1406); the curled-leaved pond-weed, *P. Crispus* (Fig. in "E. B.," 1413); the shining-leaved pond-weed, *P. lucens* (Fig. in "E. B.," 1408). All these have their leaves linear or lanceolate, and all grow under water.

The BROAD-LEAVED POND-WEED (*P. natans*)—Fig. in "E. B.," 1399—has the leaves alternate, the upper ones floating, broad and heart-shaped, with the flowering stem between; it is very common in stagnant waters and slow-running rivers. The roots are eaten with great avidity by swans. Mr. Stackhouse says: "Their love of this plant is such, that a pair of them, by harassing it in search of its succulent roots during winter, almost destroyed it in the whole extent of nearly five acres of water, which at times had been completely matted over with it." Flowers June and July. Hooker says all the species grow in water, and

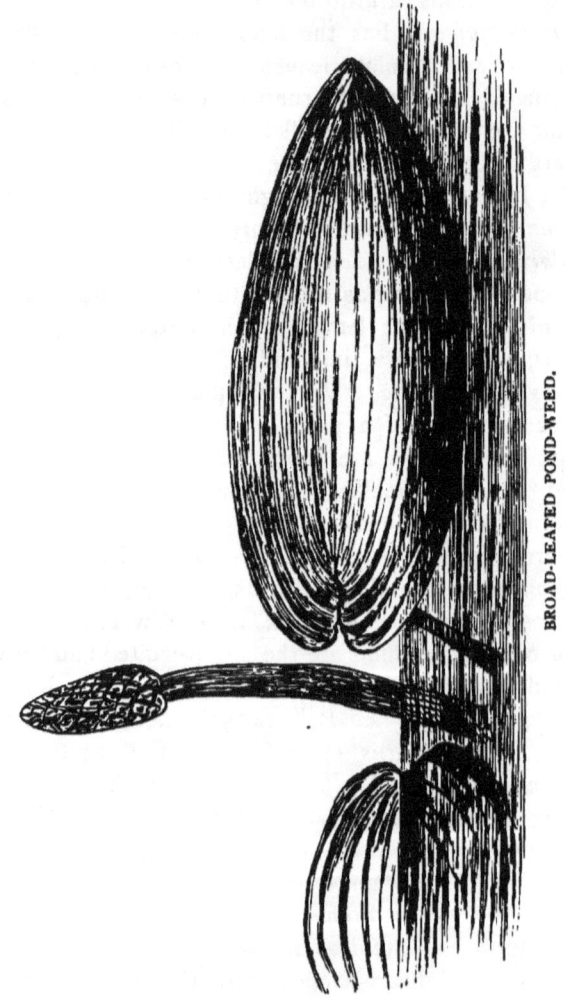

BROAD-LEAFED POND-WEED.

often present as beautiful an appearance in clear streams as the Fuci do in the ocean. They protect the spawn of fish and harbour innumerable aquatic insects, and afford food by their roots and seeds to various aquatic birds.

How well do we recollect our first sight of the FLOWER-ING RUSH or WATER-GLADIOLE! We were looking over the old bridge at Walton, and there, in the still water produced by an eddy of the river, was a lovely bunch of pink flowers. We immediately recognised our prize and made a rush to secure it. But, alas! the water was deep, and the stems were not within reach; there was nothing for it but to strip, and we soon had root, leaves, flower, and all on the dry bank. How we gloated over and examined our capture bit by bit! We have since found it in other places on our fishing excursions, but it is not a common plant. The scientific name is *Butomus umbellatus* (Fig. in " E. B.," 1443), and it is the only British plant which has nine stamens. The term Butomus is derived from the Greek words *bous*, an ox, and *tomos*, sharp, because the sharp leaves injure the mouths of cattle.

The leaves grow immediately from the root, are upright, narrow, quite entire, more or less spirally twisted at the extremity, and are two to three feet high. The large umbel of beautiful purplish pink flowers is at the end of a long flowering stalk.

"The Water Gladiole or Grassie Rush," says Gerarde, "is of all others the fairest and most pleasant to behold, and serveth well for the decking and trimming up of houses, because of the beautie and tracerie thereof." Flowers June and July.

The MARE'S TAIL (*Hippurus vulgaris*)—Fig. in " E. B.," 516—from the Greek words *hippos*, a horse, and *oura*, a tail—although not a very common plant, is not unfrequently met with in a fisherman's rambles. It is chiefly an inhabitant of stagnant waters, and is known by its erect jointed stem, with whorls of leaves at each joint, narrow and pointed, about eight in number. Flowers June and July.

Another plant, somewhat similar in appearance, but of a totally different order, is the GREAT WATER-HORSETAIL (*Equisetum fluviatile*)—Fig. in " E. B.," 1893; one of the Cryptogamia or non-flowering plants. It is frequently found in muddy lakes and by the sides of rivers and pools, the barren stems rising from three to four feet in height.

"By encouraging the growth of water-plants, which can usually be got close at hand, an increased supply of animal life is produced. It is necessary, however, to use care in the introduction of these plants, as there are some that are very objectionable. The American weed (*Anacharis alsinastrum*), for instance, that is invaluable in my ponds, where I have it thoroughly under control, is to be dreaded in all natural waters, as, once introduced, it is very difficult to get rid of, and fills a place up. The bladderwort (*Utricularia*) actually eats perch. There are many plants that are very good, however, and require but little keeping in check. Amongst those are the water-lobelia (*L. Dortmanua*) and the quillwort (*Isœtes lacustris*), which are both bottom-loving plants, and grow in from a foot to six feet of water, and form a beautiful green covering of aquatic herbage. Of the two, the lobelia prefers the shallow water, while the quillwort thrives in the deeper. The stonewort (*Chara flexilis*) is another, and I would certainly have a few white water-lilies in the pond. I have often heard anglers object to them, but they are so easily kept in check, spread so slowly, and can all be mown down bodily if desired, that I have always gone in for a few of them, and have found them very useful."

In June and July, one's attention is often attracted by the gorgeous display of the masses of yellow golden flowers of the YELLOW WATER-IRIS, CORN-FLAG, or FLEUR DE LUCE (*Iris pseudo acorus*)—Fig. in "E. B.," 1495. The exquisite colour, the beautiful curve of the flower (hence its name Iris), its tall, elegant stem, and graceful leaves, combine to make it the queen of aquatic plants. The stem rises from 2 ft. to 4 ft., somewhat zigzag, round and smooth; leaves upright, in two opposite rows, sword-shaped, ribbed, grass-green.

Curtis says: "Those who have examined the structure of the flowers of this plant must allow it to be at once beautiful, delicate, and singularly curious. The stigma, in particular, deserves to be noticed by the student, being in form and substance more like the petals than the part it really is."

The juice of the fresh root is excessively acrid. The fresh roots have been mixed with the food of swine bitten by a mad dog, and it is stated they escaped the disease, when others bitten by the same dog died raving mad.

The roots are used in Scotland to form a black dye, and are boiled with copperas to make ink. A slice of the fresh root held in the mouth between the teeth is said to cure the toothache.

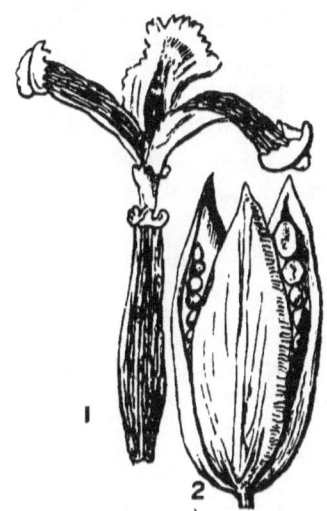

YELLOW WATER-IRIS.
1. The germen, style, and three large petal-like stigmas.
2. The capsule containing the seeds.

The Orris root of the druggist, with its pleasant violet-like smell, is really Iris root.

Another very beautiful water-plant is the FORGET-ME-NOT (*Myosotis palustris*), from *mus*, a mouse, and *ous*, *oots*, an ear, from the shape of the leaves (Fig. in " E. B.," 1104). It is also known by the names of Water-Scorpion Grass, and Marsh Mouse-Ear.

This plant at times grows in such profusion as to form a thick turquoise bed with its bright-blue flowers, varied by its golden eye and small white ray at the base of each segment. The flower-buds just before they open are of a delicate pink colour, but immediately on expansion turn blue. Flowers during the summer months.

The plant is the emblem of friendship wherever it is known. Its name, "Forget-me-not," comes from the following legend:—

Two lovers, who were on the eve of being united, were loitering on the margin of a lake on a fine summer evening, when the maiden espied a cluster of these flowers growing by the side of the bank of an island some distance from the shore. She expressed a wish to have them. Her lover plunged into the water, swam to the spot, and gathered the wished-for flowers. On returning, his strength failed him. He had only time to throw the flowers on the bank, and, as he sank, cast his eyes on the beloved of his soul, and cried, " Forget me not." (See Mills' " History of Chivalry.")

> " Little flower, whose magic name
> Kindles up affection's flame,
> Free from all the tricks of art,
> In the wayside traveller's heart,
> Pleas'd thy radiant head I view,
> Crown'd with bright cærulean blue.
> . . . . . .
> Change since then has mark'd my lot,
> Much I've seen, and much forgot ;
> Still thy pale blue light appearing,
> Childhood's earliest haunts endearing,
> Though its hours like stars have set,
> Thee and them I ne'er forget."

The COMMON BROOK LIME (*Veronica beccabunga*)—Fig. in " E. B.," 990—is frequently found in running waters and ditches, flowering through the summer months; it has beautiful racemes of many bright blue flowers, the corolla being four-cleft ; the whole plant is glabrous and very succulent.

The GIPSY WORT, WATER-HOREHOUND (*Lycopus europæus*)—Fig. in " E. B.," 1019. The name is derived from *lukos*, a wolf, and *pous*, a foot, from the fancied resemblance of the cut leaves to a wolf's paw. The stems rise often to four feet in height, and are marked with four angles. The leaves are placed opposite, close to the stem, and deeply cut. The flowers whitish, with a purple tinge, and placed in thick whorls round the upper leaves. The name Gipsy Wort arises from the gipsies using the juice of this plant to dye their skins a blackish brown. Flowers June and July.

A very beautiful plant, not uncommon in some localities, especially in the marshy meadows bordering the rivers, is the COMMON BUCK-BEAN, DOG-BEAN, or MARSH TREFOIL (*Menyanthes trifoliatum*)—Fig. in " E. B.," 920. The name is derived from the Greek *mene*, a month, and *anthos*, a flower, because the flowers are said to last in perfection for a month. The leaves are on long radical stalks, and divided into three leaflets, slightly waved at the margins. The flower stem is about eight inches in height, terminating in a raceme of flowers, flesh-coloured outside, almost white within, the disk beautifully fringed with white filaments. Flowers May and June.

> "Oft where the stream meandering glides,
> Our beauteous menyanthes hides
>   Her clustering fringed flowers ;
> Nor 'mid the garden's sheltering care,
> Of famed exotics, rich and rare,
> Purple or roseate, brown or fair,
>   A plant more lovely towers."

In the Highlands of Scotland it is used as a tea to strengthen weak stomachs. It cures the disease called the darn in cattle, and has been used as a substitute for hops. It is said to cure the sheep of the rot, but it is now ascertained that sheep will not eat it. An infusion has often been used against rheumatism and dropsy. In Hamburg it is called the " Flower of Liberty."

A handsome plant frequently found on the margin of the rivers is the COMMON MEADOW RUE or FEATHER COLUMBINE (*Thalictrum flavum*)—Fig. in " E. B.," 8—from *thallo*, to grow green. The stem grows three or four feet in height. The leaves are placed alternately—dark-green ; the leaflets are usually divided into three clefts. The flowers are very numerous—yellow—and the panicle very much branched.

Flowers June and July. The root has been used to dye wool yellow, and it has been employed as a remedy in jaundice.

The CREEPING or MARSH SPIKE RUSH (*Eleócharis palustris*)—Fig. in "E. B.," 1536—is frequently found by the sides of ditches and wet places, root creeping to a great length, black and shining, as well as the external sheaths of the stems. Flowers June and July. There are five other species of *Eleócharis*: *E. multicaulis*, very like the above, only the root is not creeping; *E. pauciflora*, rare in England, found frequently on the moors in Scotland;

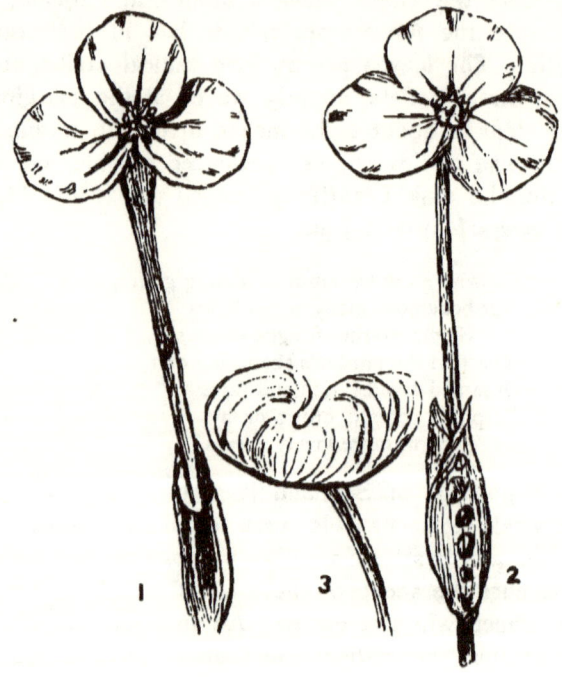

THE FROG-BIT.
1. Fertile flower.   2. Sterile flower.   3. A leaf.

E. cæspitosa and E. *acicularis*, the least spike rush, common by the sides of lakes, &c.; and *E. fluitans*, the floating rush, found in still lakes, ditches, and in places where water has dried up.

The FROG-BIT or LESSER WATER-LILY (*Hydrocharis Morsus Ranæ*)—Fig. in "E. B.," 1444—(from *udor*, water, and *charo*, to rejoice), is a very pretty plant, found in many pools and slow-running rivers. The leaves, mostly floating, are kidney-shaped, about an inch and a half broad, fleshy and smooth, somewhat transparent, and purplish underneath. The barren flowers are white, on long flowery stalks, coming out of a two-leaved transparent membraneous sheath; the fertile flowers are also white, and come out of a single sheath. Curtis says: "The whole structure and economy of the Frog-bit is exceedingly curious, deserving the minute attention of the inquisitive botanist." Flowers July.

The CLUB-RUSH or BULL-RUSH (*Scripus lacustris*)—Fig. in "E. B.," 1576. Hooker says the name is derived,

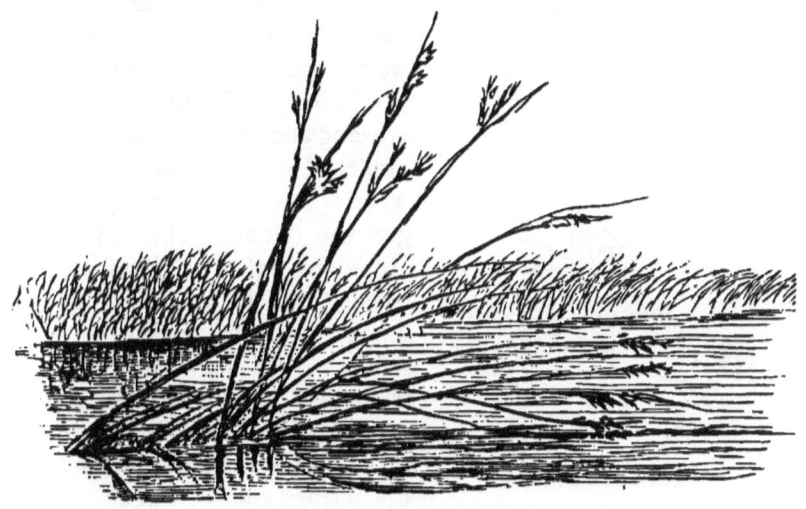

THE CLUB OR BULL RUSH.

according to Théïs, from the Celtic word *cirs*, making *cors* in the plural; whence comes the English word *cord*, the stems of this plant being made into cords—is very plentiful in our rivers, lakes, and ponds, and a very unpleasant addition it is to the other " weeds." Once catch the fly firmly in it, the chances are you leave it there. When the water is low it lies thick upon the streams. The stems are very variable in size, often as thick as a finger at the base, sometimes from 6 ft. to 8 ft. long. The flower husks (*glumes*) are fringed with brown. The stems are much used for mats, chair-bottoms, &c., and form a considerable article of trade. Coopers use them for filling up the spaces between the staves of casks. There are seven British species. Flowers July.

At the bottom of ponds and slow-flowing rivers, particularly where mud has accumulated, the COMMON HORNWORT (*Ceratophyllum demersum*) —Fig. in "E. B.," 1276— (from *keros*, a horn, and *phyllon*, a leaf, they being branched like a stag's horn), grows in considerable quantity. The stems, which float entirely under water, and at times form a portion of "the floating weeds," are long and slender, and much branched; the leaves, in whorls, are, thin, slender, and green in colour. The flowers are found in the axil of the leaves, which towards the upper extremity become very plumose. It is a very elegant plant, and affords shelter for many kinds of fish. There are two species—the above, and *C. submersum*, distinguished by the leaves being more feathery and having no spines to the fruit. Flowers July.

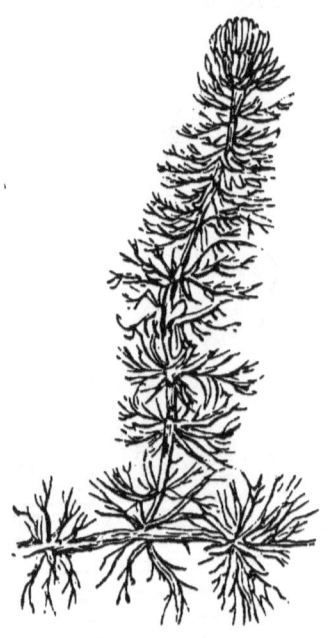

COMMON HORNWORT.

One of the commonest of our water-plants, cultivated to a very great extent as an addition to our breakfast-tables, is the WATERCRESS (*Nasturtium officinale*)—Fig. in "E. B.," 125—named from *Nasus tortus*, a convulsed nose, an effect supposed to be produced by the acid and pungent quality of this plant. The stems are spreading and creeping; leaves alternate, smooth, shiny, deep green; flowers small white, some with a purplish tinge. The wild form—which has a more pungent taste, and some think more wholesome, than the cultivated—is generally found in pebbly, gravelly streams in July.

> " Her way is o'er the dewy meads
> And by the violet dell,
> To where a plank her footsteps leads
> By the old haunted well ;
> And then she steps from stone to stone
> In the brook's gurgling water's throne,
> To where the cresses grow."

There are three other British species of this genus—*N. sylvestre*, *N. terrestre*, and *N. amphibium*, all found in watery places, the latter the most common.

Of the Bur-Reeds, *Sparganum* (from the Greek word *sparganon*, a band or ribbon, in allusion to the shape of the leaves), there are three species, distinguished by the shape of the leaves :—

1. The BRANCHED BUR-REED (*S. ramosum*)—Fig. in " E. B.," 1337—has the leaves triangular at the base, the sides concave ; found on the banks of ditches and stagnant pools.

2. The UNBRANCHED BUR-REED (*S. simplex*)—Fig. in " E. B.," 1338), leaves with flat sides ; found in pools and slow-running rivers.

3. The FLOATING BUR-REED (*S. natans*)— Fig. in " E. B.," 1339—leaves very long and pellucid ; found in lakes and stagnant waters.

All three species prefer a gravelly to a muddy bed. Flowers July.

The WATER-FIGWORT or WATER-BETONY (*Scrophularia aquatica*)—Fig. in " E. B.," 947—(from *scrofula*, a disease which this plant is supposed to cure). Somehow or other the plant always pokes its branched head up just where you do not want it, and seems to take a pleasure in twisting your line should you unfortunately get caught in it, particularly after the efflorescence, when the seeds are getting ripe. The stems rise to four feet in height. The flower panicles terminal and branched. Flowers dark purple at the mouth, with a scale on the upper lip. Flowers July.

A very beautiful but rare plant grows in the upper part of the Thames, the Isis, and Cherwell, and in some of the slow-running rivers of Cambridgeshire and Norfolk, as well as in ornamental waters—the FRINGED WATER-LILY (*Villarsia nymphæoides*)—Fig. in " E. B.," 921. The flowers are large and yellow and very curiously plaited. Leaves green, heart-shaped, floating. Flowers June to September.

The canals in Holland are often covered with this flower, and, according to Kæmpfer, the Japanese salt the leaves, which, by this process, become very glutinous, and are used in soup.

Of the Water-Dropworts (*Œnanthe*), (from the Greek *oine*, a vine, and *anthos*, a flower, alluding to the vinous smell of the blossoms), the three most common species are :—

1. The COMMON DROPWORT (*Œnanthe fistulosa*)—Fig. in " E. B.," 593. Plant two to three feet high, chiefly found in ditches and rivulets. Flowers July and August.

2. The HEMLOCK WATER-DROPWORT (*Œnanthe crocata*) —Fig. in " E. B.," 597. Plant three to four feet high, differing from the other species in the great breadth of its leaflets, and large, much ramified stems, full of a very poisonous yellow juice. Flowers white. This is one of the most poisonous of the umbelliferous plants, and as the root is like celery, some very serious and fatal mistakes have occurred. It flowers in June and July, and in some places grows in great abundance.

3. The FINE-LEAVED WATER-DROPWORT (*Œnanthe phellandrium*)—Fig. in " E. B.," 598. Flowers July. All these, if allowed, grow to a considerable height, and we know of nothing more tiresome than the line getting entangled in a branch of Water-Hemlock.

The COMMON YELLOW WATER-LILY (*Nuphar lutea*)—Fig. in " E. B.," 54; from *Naufar*, Arabic for *Nymphæa*), also called Yellow Watercan, is a beautiful and elegant ornament in many of our lakes and slow-running rivers. Its large floating leaves—some of them a foot wide—and bright yellow flowers are very conspicuous. The flowers smell like brandy, and from this and the shape of the seed-vessels, this plant in some localities goes by the

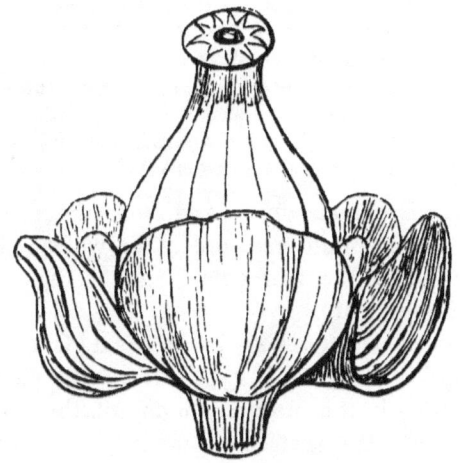

SEED-VESSEL OF NUPHAR LUTEA.

provincial name of Brandy Bottles. The roots bruised in milk are said to be poisonous to beetles and cockroaches.

THE LARGE WHITE WATER-LILY (*Nymphæa alba*)—Fig. in " E. B.," 53—is another beautiful object.

> "The water-lily to the light
> Her chalice rears of silver bright."

It is very common in the Isis and Cherwell.

"Of all our native plants, the White Water-Lily is the most magnificent; in size, beauty, and elegance of its corolla, it may vie with many of the finest magnolias of America; and its delicate and pure white petals are little inferior to those of the night-blowing Cereus."

Tom Moore says—

> "The virgin lilies all the night,
> Bathing their beauties in the lakes,
> That they might rise more fresh and bright
> When their beloved sun awakes."

And Mrs. Hemans in " National Lyrics "—

> " Come away, elves, while the dew is sweet ;
> Come to the dingles, where fairies meet.

> Know that the lilies have spread their bells
> O'er all the pools in our forest dells;
> Stilly and lightly their vases rest
> On the quivering sleep of the water's breast,
> Catching the sunshine thro' leaves that throw
> To their scented bosoms an emerald glow;
> And a star from the depth of each pearly cup—
> A golden star—unto heaven looks up,
> As if seeking its kindred, where bright they lie,
> Set in the blue of the summer sky."

The flower closes at sunset, and from the stem sinking a little, the flower often lies under the water; it expands in the daytime, as the sun gets power. Fig. in "E. B." 160. Flowers in July.

The COMMON DUCKWEED—Duck Meat, Greeds (*Lemna minor*)—Fig. in "E. B.," 1395—from the Greek word *lemna*, a scale—is the most abundant of the four British species, covering the surface of many ponds and ditches, harbouring numerous insects and molluscæ, the food for ducks and other waterfowl. If you take some of it up in your hands, you will see how delicate the fronds are, and notice the long solitary root hanging down. Flowers in July.

The GREAT WATER-DOCK (*Rumex hydrolapathum*)—Fig. in "E. B.," 1220—grows abundantly on the banks of our rivers, and sometimes the fly-fisher's temper is slightly ruffled by finding his line entangled in the stems of this plant. The stems are from three to five feet high, and some of the lower leaves measure one and a half feet. It is when seeding that it is particularly apt to catch the line. Flowers July and August.

The PURPLE LOOSESTRIFE (*Lythrum salicaria*)—Fig. in "E. B.," 491—with its elegant spike of purple flowers, is one of the most conspicuous features on the banks of our rivers and ornamental waters and lakes. The stem rises from three to five feet, upright, four-angled, tinged with red, and sometimes downy. The leaves are green above, lighter underneath. Flowers in tufts of a beautiful crimsony purple. Flowers June to September. A strong decoction of this plant is often used by the country people for the cure of dysentery and diarrhœa.

This plant must not be confounded with the LOOSESTRIFE (*Lysimachia vulgaris*)—Fig. in " E. B.," 1141—which is often found on the banks of rivers, with its erect two or three feet stem, glabrous leaves, and handsome yellow flowers. Flowers July. Pliny says this plant has the power to tame restive horses.

The MEADOW SWEET or DROPWORT (*Spiræa ulmaria*)—Fig. in " E. B.," 415—is a great favourite with those who wander by the river-side. It attracts immediate attention by its tall reddish stems, and waves of yellowish-white sweet-scented flowers, and its handsome jagged leaves. The French call it *La Reine des Prés*, and it well deserves the name. The stem rises from three to five feet high, branched upwards; the leaves are downy beneath, serrated, large, the terminal one the largest. The flowers are yellowish-white, and very sweet-scented. Flowers July.

Of the WILLOW HERBS (*Epilobium*)—from *epi*, upon; *lobos*, a pod; the flower being placed on the top of the seed-vessel, four species are found along the margins of our waters, all having purple flowers, and when seeding, from the height of the flowering stem, it is occasionally annoying to fly-fishers.

The ROSE BAY WILLOW HERB (*Epilobium angustifolium*)—Fig. in " E. B.," 495—is the largest; it is rare in England, but common in Scotland; distinguished from the others by the flowers being irregular and the stems bent down, whilst the two following have the flowers regular and the stems erect. The Rose Bay is a very handsome plant, four to six feet high. Flowers July.

The GREAT HAIRY WILLOW HERB (*E. hirsutum*)—Fig. in " E. B.," 497—is almost as large; common enough on our rivers and lakes; stems from four to five feet high, much branched; flowers large and purple. Flowers July.

The SMALL FLOWERED WILLOW HERB (*E. parsiflorum*)—Fig. in " E. B.," 493—similar to the above, but much smaller in all its parts; stems from one to one and a half feet high, which distinguishes it. Flowers July. The flowers of the Willow Herbs have a peculiar odour, likened by some to that of a gooseberry-pie.

The Common Reed (*Arundo phragmitis*)—Fig. in "E. B.," 1727—so abundant on the margins of our lakes and rivers, is one of the tallest and handsomest of our grasses, runs often to six feet or more in height. The panicle (flowers) large purple-brown, at first upright, afterwards spreading, and finally drooping. The stems (culms, as they are called) upright, stout, from five to seven feet high, hollow, with

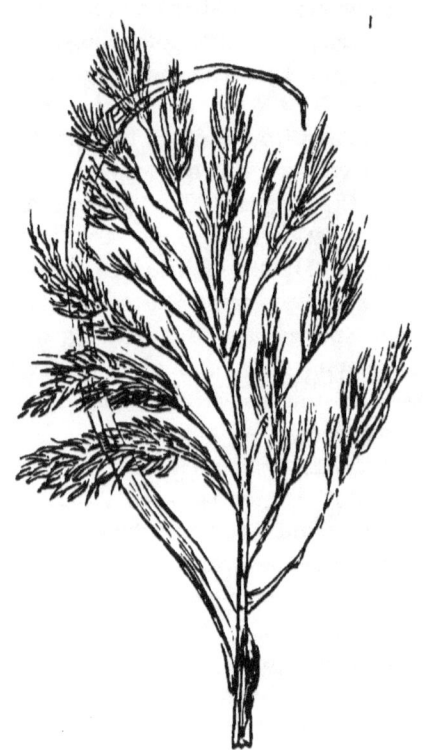

THE COMMON REED.

many knots; very smooth, shiny, and leafy. Leaves a foot long or more, spear-shaped, broad, many ribbed, tapering to a very fine point at the apex. The spikelets dark purple, from three to six flowered; as the flowers advance the tufts of hair increase, at length becoming very silky. This plant frequently forms patches of immense extent, called reed-ronds in some parts of the East of England, which harbour many birds, amongst them the Bearded Titmouse (*Parus biamarcus*) and the Reed-Warbler. An extensive use is made of the culms for thatching, &c. In Sweden it is used as a dye. The creeping stems contain sugar, and the young shoots make a good pickle. The stem was formerly used as a pen (*calamus*) for writing. Flowers July. The common fishing-rods of the Continent, particularly in 'France, Switzerland, and Italy, are fabricated of the much stouter stems of the *Arundo donax*, a native of the South of Europe.

Of the same order is the REED CANARY-GRASS (*Phalaris arundinacea*)—Fig. in "E. B.," 1697—which is often found by the sides of rivers and lakes. It is used much for securing river-banks. It is cultivated in gardens, and has variegated leaves; goes by the name of Ribbon-Grass. Flowers July and August.

The REED MEADOW-GRASS (*Poa aquatica*)—Fig. in "E. B.," 1751—rises four to six feet in height. The panicle of flowers much branched and erect; leaves sword-shaped. Flowers July and August.

The FLOATING MEADOW-GRASS (*Poa fluitans*)—Fig. in "E. B.," 1752–1753—another of our river grasses, with creeping roots; gives shelter to fish and to the various larvæ of the water-insects. Flowers July and August.

The GREAT CAT'S TAIL or REED-MACE (*Typha latifolia*)—Fig. in "E. B.," 1685—from τιφος, a marsh, improperly called the Bull-Rush, which is really a totally different plant (see *Scirpus lacustris*); is frequently found on the borders of rivers and lakes, associated with the Yellow Iris and Common Reed. The stems rise from three to six feet high; leaves very long, sometimes nearly

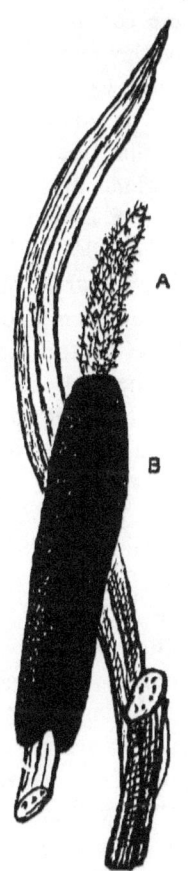

THE REED-MACE.

an inch broad, are used by coopers; catkins very long and close together, the fertile (B) greenish-brown, the sterile (A) yellow, with one or two large membranaceous bracteas. Flowers July and August. The pollen of the sterile flowers is very inflammable, and is used by firework-makers instead of the Club Moss. Flowers July and August.

There is another species, the LESSER CAT'S TAIL or REED-MACE (*T. angustifolia*)—Fig. in "E. B.," 1336. Much less common, easily distinguished from the Great Cat's Tail by the leaves being linear and *grooved below;* sterile and fertile catkins a *little distant from each other.* Flowers July.

The young shoots of both these species are much eaten by the Cossacks of the Don, and are sometimes used in England under the name of Cossack Asparagus.

The MARSHWORT or FOOL'S WATERCRESS (*Helosciadium nodiflorum*)—Fig. in "E. B.," 573–574—from the Greek word *elos*, a marsh, and *skiadion*, an umbel, is found by the side of our lakes and small rivers. The leaves greatly resemble those of the Common Watercress, but are readily distinguished by the dilated sheathing base of the leaf-stalk, which is not so in the true Watercress. The juice mixed with milk is a popular remedy in some cutaneous disorders. The flowers are small and white. Flowers July and August.

The WATER-HEMLOCK or COWBANE (*Cicuta virosa*)—Fig. in "E. B.," 571. Hooker says Cicuta was a term given by the Latins to those spaces between the joints of a reed of which their pipes were made, and the stem of this plant is similarly marked by hollow articulations. Although not very common, it occasionally is found by the side of some of our rivers. We have found it on the Itchen. The juice is a *deadly poison*, very fatal to man and horned cattle, yet goats devour it eagerly, and sheep and horses eat it with impunity. The stem rises from two to four feet high, hollow and leafy; the leaves are on long stalks, bright green; leaflets pointed and spear-shaped; the clusters of flowers white, upright, and large. Flowers July and August.

The BROAD-LEAVED WATER-PARSNEP (*Sium latifolium*)—

Fig. in "E. B.," 537—rears its umbelliferous head in slow-running streams in July, when the so-called weeds are allowed to grow. (The NARROW-LEAVED WATER-PARSNEP (*Sium angustifolium*), is also common enough, but does not grow so high.) The stem is sometimes as high as six feet, often from four to five; it is deeply furrowed and leafy; the leaflets are spear-shaped and serrated; those of the leaves which grow under water are often pinnatifid (feather-cut). Hooker says the *Sium* is derived from the Celtic word, *siw*, water. Flowers July and August.

The COMMON ARROW-HEAD (*Sagittaria sagittifolia*)—Fig. in "E. B.," 1436—is another of the beautiful aquatics found on the margin of our rivers in the months of July and August. The leaves come from the root on long, triangular, very cellular foot-stalks. Those below the surface are long and strap-shaped. Those above the water are large and arrow-shaped, very entire, with parallel ribs and reticulated veins. The flowers, three in each whorl, are white, with a purplish tinge at the claw.

The GREAT WATER-PLANTAIN, GREAT THRUMWORT (*Alisma plantago*)—Fig. in "E. B.," 1438—from the Celtic word, *alis*, water, is another very common plant. The stem is from two to three feet high. The leaves all start from the root on long stalks. Flowers of a pale rose colour. This plant was supposed to be a cure for hydrophobia, and was the chief ingredient in all the drinks given for that purpose; it was also supplied as poultices to the bite. Flowers July and August.

There are two other species: the FLOATING WATER-PLANTAIN (*A. natans*)—Fig. in "E. B.," 1441—found on the lakes in North Wales and Cumberland; and the LESSER WATER-PLANTAIN (*A. ranunculoides*), found in ditches and wet places.

In large open waters or in ponds may be noticed a longish, thread-like, much branched, floating weed, with pretty-looking white star-like flowers, one (the fertile) with a long anther. This is the COMMON HORN POND-WEED (*Zannichellia palustris*)—Fig. in "E. B.," 1425–1426; Nat. Ord. *Fluviales*)—named, according to Hooker, in honour of

John Jerome Zannichelli, a Venetian apothecary and botanist. Flowers August.

The HEMP or WATER-AGRIMONY (*Eupatorium cannabinum*) —Fig. in " E. B.," 785—is common enough on the banks of rivers and other watery places. Stems from three to four feet high, very tough and unwieldy, leaves downy. Three to five partite, middle lobe the longest and opposite. Flowers very numerous, of a pale reddish-purple, thickly crowded into tufts, terminating the stem and upper branches. Flowers July and August.

The COMMON FLEA-BANE or HERB CHRISTOPHER (*Pulicaria dysenterica*)—Fig. in " E. B.," 770—from *pulex*, a flea ; the powerful smell being obnoxious to fleas. The stem rarely exceeds two feet in height, is firm, solid, and cottony, more or less branched. Leaves spreading, dull green above, whitish underneath, and cottony. Flowers yellow at the top of the stem, two or three together. Flowers July to October.

It is said that the Russian soldiers, in their expedition to Persia under General Keil, were cured of dysentery by this plant; whence Linnæus gave it the specific name of *Dysenterica*.

There is another plant which goes by the name of FLEA-BANE (*Erigeron*), but this never grows near the water.

The MARSH WOUNDWORT or ALL-HEAL (*Stachys palustris*) —Fig. in " E. B.," 1669—named from σταχυς, a spike, from the form of the inflorescence. The English names *Woundwort*, *All-Heal*, arise from its high repute in former times as a vulnerary. The stems rise about two feet in height. The leaves are placed opposite each other, and close to the stem ; silky above, downy beneath. The flowers are of pale purple variegated with violet and white, in whorls on the spike, with eight or ten flowers in each whorl. Flowers August.

Some of the Mints are also found by the margin of rivers and lakes, particularly the TALL RED MINT (*Mentha rubra*) —Fig. in " E. B.," 1033. The stem rises at times as high as from four to five feet. The flowers are purplish-red, the leaves are stalked and smooth. Flowers September.

The HAIRY MINT (*M. hirsuta*)—Fig. in " E. B.," 1030—is a very common plant. The stem rises to three feet. The leaves are green and hairy. The flowers are pale purple in whorls. The flower-stalks are densely covered, particularly at their summits, with recurved, at times closely pressed, white hairs. Flowers August and September. Tea made of the green leaves of this species is excellent in all nervous and hysterical cases. Mice have a particular aversion to the smell of Mint.

A water-plant has been unfortunately introduced into this country which has increased to such an extent in many localities as to completely block up the waterway in many canals and slow-running rivers. This is the WATER-THYME (*Anacharis alsinastrum*)—Fig. in " E. B.," 1446—*Elodea Canadensis* of Sowerby and other botanists, a native of North America, supposed to be imported with some timber, and first noticed in 1842 in the lake at Dunse Castle, Berwickshire, by the late Dr. Johnston, and in 1847 by Miss Kirby in the reservoirs of a canal in Leicestershire; then in the river White Adder, and by degrees extending itself in all directions. It is a dark-green, much-branched perennial, entirely floating under water, its flowers only appearing above water for a short time at the period of fertilisation. It has numerous leaves, which are either opposite or in whorls of three or four without fore-stalks. Its rapidity of growth is extraordinary. Immense masses disfigure the shallows of the Trent and cover the bed of the deeps. The stems are very brittle, and every fragment is capable of growing. Waterfowl are very fond of it, and by these means the seeds may be carried from one piece of water to another. It is said that in some instances it has been in such profusion and thickness as to impede the ascent of the salmon; but it certainly affords a shelter of safety to many other kinds of fish. Fortunately swans are extremely fond of it, and where it abounds it can be almost eradicated by placing a pair of swans on the water. In the *Fishing Gazette* appeared the following paragraph :—

" According to Dr. Brandes, of Hitzackes, Hanover, the American weed *Anacharis alsinastrum* is, after all, an angel in disguise. In the district where Dr. Brandes lives, he has noticed that malaria and diarrhœa, which used to appear annually in a sporadic or epidemic form, commenced to decrease very soon after the American weed made its appearance, and during the last four years have been altogether wanting. Dr. Brandes is of opinion that the weed nourishes itself to a great extent on decayed vegetable matter, and where this is abundant it grows with great rapidity. It thus destroys the germs of certain diseases. As manure, the weed is said to be excellent. In some waters it seems to have grown itself out, and is gradually disappearing. If Dr. Brandes is correct, the probable cause of its disappearance is that it has fulfilled its mission, purified the river, and died for lack of further impurities to feed on."

Such are the principal plants found either in the water, on the banks, or in the moist meadows and marshes.

> " Beautiful things ye are, where'er ye grow :
> The wild red rose, the speedwell's peeping eyes,
> Our own bluebell, the daisy, that does rise
> Wherever sunbeams fall or winds do blow ;
> And thousands more, of blessed forms and dyes :
> I love ye all."

INDEX.

ABERDEVINE or siskin, 120.
Adder, the, 169-171.
Alder-fly, the, 319.
Alder-moth, the, 353.
All-heal, the, or marsh woundwort, 392.
"Amateur Angler," the, and the devil's coach-horse, 304-5.
Anderson, Mr., and grilse, 249.
Andrews, Mr., and Loch Leven trout, 266.
Ants, the, 347-8.
Archer, Mr. W. C., and salmon, 244.
Argyll, Duke of, and the water-ouzel, 39.
Armistead, Mr., and the char, 276.
Arrow-head, the common, 391.
Azurine, the, 226-7.

BAËTIS, the genus, 340-6.
Bailey and the bearded titmouse, 30.
Baker, Sir Richard, and the carp, 206.
Barbastelle bat, 19.
Barbel, the, 209-11.
Barker ("Art of Angling") and the pike, 285.
Barn-owl, the, 157-9
Bartlett, Mr., and the water-ouzel, 40-2.
Basse, the, 198-9.
Bats, the, 17-20.
Bearded titmouse, the, 30.
Beart, Mr. W. F., and the spinner, 329.
Beetles, the (or Coleoptera), 301-8.
Bell, Mr. ("British Quadrupeds and Reptiles"), and the field-vole, 5; and the otter, 9; and the stoat, 13-14; and shrew-mouse, 15; and water-shrew, 16-17; and bats, 18-19; and the lizard, 168; and the slow-worm, 168-9; and the frog, 171-4; and the toad, 175.
Benson, Mr., and the martins, 78.
Bewick and the teal, 55.
Birds, the, 21-165 (see Kingfisher, &c., &c.).
"Birds of Europe," see Dresser.
Bittern, the, 63-5.
Blackbird, the, 94-6.
Blackcap, the, 108-10.
Black gnat, the, 352.
Bladderwort, the great, 370-3.

Blaine's "Encyclopædia of Rural Sports," 178; and perch, 197; and ruffe, 198.
Blandford, Mr., and the cuckoo, 92.
Bleak, the, 227-8.
Blethisa multipunctata, the, 307.
Bloomfield and the white-throat, 102.
Blue-bell, the, 363.
Bluebottle, the, 352.
Blue titmouse, the, 134-5.
Boat-fly, the, 302-3.
Bog-asphodel, the, 364.
Bog-bean, the, 379.
Brander, Mr., and the salmon, 246.
Brandes, Dr., and the water-thyme, 393-4.
Bream, the, or carp-bream, 219-22.
"British Quadrupeds," see Bell.
Broderip's "Zoological Recreations," 15-16.
Brook-lime, the common, 378.
Brown, Mr., and salmon smolts, 242, 248-9.
Browne, Mr., and the perch, 195; and the carp, 207; and the tench, 216; and the roach, 223; and the rudd, 225; and the pike, 281.
Browne, Sir Thos., and the coot, 35.
Browning and the robin, 105.
Buck-bean, the common, 379.
Buckland, Frank, and the thrush, 94; and the curlew, 152; and the owl, 158; and perch, 197; and the stickleback, 202; and the carp, 208; and the barbel, 209; and the tench, 216; and the bream, 220; and the roach, 223; and the rudd, 225; and the dace, 226; and the twaite-shad, 230; and salmon smolts, 242; and the bull-trout, 261; and river-trout, 268-9; and the pike, 283; and eels, 287-8.
Bull-rush, the, 381-, 2
Bull-rush moth, the, 353-4.
Bull-trout, the, 261-2.
Bund, Mr. Willis ("Salmon Problems"), 239, 241, 245, 251; and eels, 290.
Bunting of the bulrushes, see Reed-sparrow.

INDEX.

Burbot, the, 186.
Bur-marigold, the, 367.
Bur-reeds, the, 383.
Butter-bur, the common, 364-5.
Buttercup, the, 360.
Butterflies and moths, 352-5.

CAREW, THOS., and the cuckoo, 90.
Carex, or the great sedge, 365.
Carp, the, 206-9, 218-9.
Carp-bream, the, 219-22.
Carthusian Convent and the otter, 5-6.
Cat's tail, the great and lesser, 389-90.
Chaffinch, the, 119-20.
Char, the, 274-6.
Chaucer, and the bittern, 64; and the skylark, 80; and the sparrow-hawk, 154; and the pike, 284.
Chiff-chaff, the, 100-1.
Chub, the, 223-4.
Clare and the moor-hen, 33; and the water-cups, 366.
Cloëon, the genus, 340-6.
Club-rush, or bull-rush, 381-2.
"Coachman," the, 353.
Coal-titmouse, or the cole-tit, 130-2.
Cockroach, the, 323.
Cock-tail beetle, 303-5.
Collier, Mr., and the otter, 6-7.
Comfrey, the, 367.
Common frog, the, 171.
Common lizard, the, 168.
Common loach, the, 213.
Common sandpiper, the, 65-6.
Common toad, the, 174-5.
Common viper, the, 169-71.
"Compleat Angler," the, *see* Walton.
Coot, the, 33-6.
Corn-flag, the, 376-7.
Cotton and the grayling, 272.
Cowbane, the, or water-hemlock, 390.
Cow-dung fly, the, 351.
Cowper's poem on the snail, 296.
Cowslip, the, or paigle, 361-2.
Crake, W. P., and the May-fly, 330-1, 335.
Cray-fish, the, 296-7.
Creeping or marsh spike-rush, the, 380.
Cricket, the common, 323.
—— the water, 349-50.
Crow, the, or rook, 136-46.
Crowfoot, the celery-leaved, 368.
—— the upright meadow, 360.
—— the water, 367.
Crucian carp, the, 218-19.
Cuckoo, the, 86-92.
Curlew, the, 151-3.
Curtis, and the buttercup, 360; and the water-iris, 376.
Cuvier on fishes, 179; and the perch, 196-7.

DABCHICK, the, 44-6.
Dace, the, 225-7.
Daffodil, the common, 363.

Daisy, the, 360-1.
Daniel, and the perch, 195; and the ruffe, 198; and the carp, 206; and the gudgeon, 212; and the tench, 216-18.
Darwin ("Origin of Species"), and Salmonidæ, 232-3; and earth-worms, 356-7; and the bladderwort, 371-3.
Daubenton's bat, 19.
Davy, Sir H., and the swallow, 73-4; and the magpie, 148; on fishes, 191-2; on "Salmonia," 270.
—— Dr. John, on fishes, 189; and the char, 275.
Day, Dr., and the perch, 193; and the pope, 198; and the sticklebacks, 201-3; and the Physostomi, 204-5; and the carp, 208-9; and the barbel, 210; and the gudgeon, 212; and the loaches, 213; and the tench, 216; and the bream, 220; and the dace, 226; and the twaite-shad, 230; and Salmonidæ, 233, 235, 240-2, 247, 249, 250, 252; and the sea-trout, 259-61; and the river-trout, 262-70; and the char, 274-5; and the smelt, 277; and the pike, 282; and the eels, 286-9; and the lampern, 292.
Devil's coach-horse, the, or cock-tail beetle, 303-5.
Dipper, the, or water-ouzel, 38-44.
Diptera, the order, 351.
Dishwasher, *see* Water-wagtail.
Diving-goose, *see* Merganser.
Dog-violet, the, 363.
Dor-beetle, the common, 305.
Dor-hawk, the, or night-jar, 161-5.
Dotterel, the, 150-1.
Down-hill fly, the, 352.
Dragon-fly, the great, 310-15.
Drayton, and the water-hen, 33; and the coot, 34; and the dabchick, 46; and the snipe, 60; and the heron, 62; and the bittern, 65; and the skylark, 80; and the blackbird, 95; and the hedge-sparrow, 114; and the dotterel, 151; and the owl, 158.
Dresser ("Birds of Europe"), and the reed-warbler, 28; and the bearded titmouse, 30; and the coot, 35; and the sand-piper, 66; and the martin, 78; and the cuckoo, 90; and the wry-neck, 123; and the titmouse, 133; and the owl, 159; and the night-jar, 163.
Drinker-moth, 353.
Drop-wort, the common, 384; or meadow-sweet, 387.
Dryden and the bittern, 65; and the martin, 78.
Duckweed, the common, 386.

EAGLE, the golden, 156.
Earth-worms, the, 356-8.

INDEX. 397

Eaton, Rev. A. E., and the Ephemeridæ, 324, 327-45.
Edible frog, the, 171.
Edward, Mr. Thomas, and the water-ouzel, 43.
Eels, the, 286-91.
Elliot and the skylark, 80.
Encyclopædia Britannica and insects, 314-15, 324.
Ephemeridæ, the, 324-46.
Ermine weasel or stoat, 12-14.
Erskine, Lord, and the rook, 144.

FABER, FATHER, and the nightingale, 98; and the wheat-ear, 110-11.
Fallow-chat, the, 110-12.
Feather-columbine, the, 379.
Fern-fly, the, 307.
Field, The, and otter, 7-9; and water-ouzel, 40-2; and the rook, 138; and the salmon, 247-8, 254; and the spinner, 329.
Field-mouse, the short-tailed, 5.
Field-vole, the red, 5.
Fire-tail, the, 113-14.
Fisherman's curse, the, 352.
Fishes, the (*see* Perch, Salmon, &c., &c.), 178-292.
Fishing Gazette, and water-vole, 2; and the otter, 9; and dabchick, 44; and "Do fish feel pain on being hooked?" 190; and perch ova, 196; and salmon, 246; and Loch Leven trout, 266; and the char 276; and water-thyme, 393-4.
Flea-bane, the common, 392.
Flies, gnats, &c. 351-2.
Flowering-rush, the, 375.
Flowers, the river-side, 359-94.
Fool's watercress, the, 390.
Forget-me-not, the (water-plant), 377-8.
Forster ("Scientific Angler") on fishes, 183.
Fox-moth, the, 353.
Francis, F., and the perch, 195-6; and the roach, 222 ; and the sea-trout, 260.
" Fresh-Water Fishes of Europe," 243.
Frog-bit, the, 380-1.
Frog-hopper, the, 350.
Frogs, the, 171-4.

GADOW, Professor, and the salmon, 247.
Garrick, Mrs., and the carp, 207-8.
Gascoigne, Geo., and the nightingale, 96-7.
Gawky, the, or cuckoo, 86-92.
George, Mr., and the otter, 7-8.
Gesner and the barbel, 209.
Gillaroo trout, the, 269.
Gipsywort, the, 378-9.
Gnat, the black, 352.
Golden dun midge, the, 351.
Golden eagle, the, 156.
Goldfinch, the, 117-18.

Goldsmith and the bittern, 64.
Goode, Dr. Brown, and sea-trout, 276.
Goosander, the, 54.
Gosden, Mr., and salmon, 253-4.
Gossamer webs, 354-5.
Gould and the titmouse, 130.
Governor-fly, the, 349.
Graham and the skylark, 83-4; and the goldfinch, 118; and the chaffinch, 119.
Grannom, the, 318.
Grasshopper, the, 322-3.
Grass of Parnassus, 364.
Gravel-bed or spider-fly, 352.
Grayling, the, 271-4.
Great bat, the, 18.
Great spotted woodpecker, the, 127-8.
Great titmouse, the, 133.
Green woodpecker, the, 124-7.
Gudgeon, the, 211-12.
Guillim's "English Heraldry," 22.
Günther on fishes, 183, 186, 204-5; and perches, 193; and the barbel, 209; and the tench, 218; and the Salmonidæ, 233-4, 237, 241 ; and sea-trout, 261 ; and river-trout, 262-9 ; and the char, 275; and the pike, 280; and the eel, 286; and the lamprey, 291.
Gyrinus natator or whirligig, 301-2.

HAIRY violet, the, 363.
Halford, Mr. F. M., on the Ephemeridæ, 339.
Hall, Mr. S. C., and the wren, 103-4.
Hancock, Mr., and the snipe, 59-60.
Harper, Mr. A., and mergansers, 54.
Harting, Mr., and the heron, 63; and the thrush, 94; and the night-jar, 163.
Hatton, Sir C., and the swan, 51.
Hawker, Col., and the coot, 34.
Hedge-sparrow, the, 114-16.
Hemans, Mrs., and the swan, 52; and water-lilies, 385-6.
Hemiptera, the order, 349-52.
Hemp or water-agrimony, the, 392.
Henslow, Prof., and the cowslip, 361.
Hepell and the carp, 207.
Herb christopher, the, 392.
Heron, the, 60-3.
Herrick, R., and the robin, 106.
Heysham, Mr., and the dotterel, 151.
Heywood, J., and the cuckoo, 88-9.
Hobby, the, 153.
Hoffmeister, and worms, 357.
Holinshed ("Chronicles, England"), and salmon, 257-8 ; and the pike, 284.
Hooker, Dr., "British Flora," 359 ; and the pond-weed, 374; and the bullrush, 381.
Horn pond-weed, the common, 391-2.
Hornwort, the common, 382.
Horse-leech, the, 358.
House-fly, the common, 352.

House-sparrow, the, 121.
Hunter, John, and the nightingale, 96.
Huxley, Prof., and the water-vole, 1-2; and Salmonidæ, 233; and the crayfish, 297.
Hyacinth, the wild, 363.
Hymenoptera, the, 347-55.

INSECTS, the, 299-355.
Iris, the yellow water, 376-7.
Izaak Walton's "Compleat Angler," *see* Walton.

JACKDAW, the, 147-8.
Jardine, Sir W., and the water-ouzel, 40.
Jay, the, 148.
Jesse, Mr., and moor-hens, 32; and wild-ducks, 48; and the roach, 222.
Johns ("British Birds") and the water-wagtail, 66.
Johnstone, Mr., and salmon and grilse, 250.
Jones, Prof. R., on fishes, 179-82, 191.
Jonson, Ben, and the toad, 175.

KEATS and the minnows, 229.
Kelson, Mr. G. M., on salmon, 183-4.
Kent, Miss, and the daisy, 360-1.
Kerr, Mr., and bull-trout, 261.
Kestrel hawk, the, 155-6.
Kingfisher, the, 22-5.
Kirby and the dragon-fly, 312-14, 317-18; and ants, 347-8.
Knox and the sparrow-hawk, 154.

LAMPREYS, the, 193, 291-2.
Land and Water, and salmon smolt, 242-3; and salmon, 253.
Landmark, Prof. A., and salmon-leaps, 255-6.
Lapwing, the, 56-8.
Leech, the, 358.
Lepidoptera, the order, 352-5.
Lesser-spotted woodpecker, the, 128.
Leyden and the daisy, 361.
Libellula puella, the 312-13.
Lilford, Lord, and the coot, 35; and the water-rail, 37-8.
Linnæus and the perch, 195; and Salmonidæ, 232; and the term larva, 300.
Linnet, the, 120.
Little grebe, the, or dabchick, 44-6.
Lizards, the, 166-8.
Lloyd, Mr., and the grayling, 272.
Loaches, the, 213-15.
Logan's lines about the cuckoo, 92.
Long-eared bat, 19.
Long-tailed titmouse, 132-3.
Loosestrife, the purple, 386-7.
Lubbock, Sir J., and the bream, 221; and the Cloëon, 325; and the nymph, 330.

M'BEAN, Mr., and the rook, 139-41.

Macgillivray, and the dipper, 39; and the cuckoo, 89-90; and the thrush, 94; and the nightjar, 163.
M'Gregor and the redbreast, 107.
Magpie, the, 148.
Mallard, the, 46-9.
Manley, Mr., and "Notes on Fish and Fishing," 189-91; and the perch, 196-7; and the gudgeon, 212; and the rudd, 225.
Mare's tail, the, 375.
Marigold, the bur, 368.
——— the common marsh, 364.
Marlow-buzz, the, 307.
Marryat, Horace, and the cuckoo, 89.
Marshall, Mr. P., and the salmon, 237.
Marsh penny-wort, the, 365-6.
Marsh spike-rush, the, 380.
Marsh titmouse, the, 129-30.
Marsh trefoil, 379.
Marsh-wort, the, 390.
Marsh-woundwort, the, 392.
Martin, the, 77-9.
Marvel, A., and the thrush, 94.
Matchell, Rev. J. C., and how to cook a swan, 53.
May-bug, the, 307.
May-fly, the, 328-39.
Meadow-grass, the, 389.
Meadow-pipit or titlark, 84.
Meadow-rue, the common, 379.
Merganser, the red-breasted, 53-4.
Merle, *see* Blackbird.
Merlin, the, 153.
Metzger, Prof., and the water-ouzel, 40.
Meyer, Mr. W., and the snipe, 59.
Milfoil, the spiked water, 364.
Miller's thumb, the, 200-1.
Mill's "History of Chivalry," 378.
Milman, Rev. H., and the skylark, 82-3.
Milton ("Paradise Lost"), and the swan, 50; ("Il Penseroso") and the nightingale, 97; and the cowslip, 362.
Minnow, the, 228-9.
Mints, the, 392-3.
Mire-drum, *see* Bittern.
Mitchell, Mr. A., and the salmon, 237.
Molluscs, Crustaceans, &c., 293-8.
Montgomery and the blackbird, 96; and the cowslip, 362.
Moore, Tom, and water-lilies, 385.
Moor-hen, the, 31-3.
More, H., and gossamer web, 355.
Moths and butterflies, 352-5.
Müller, Herr, and the water-ouzel, 40; and the cuckoo, 90; on fishes, 185-8.

"NAHANITE" and the water-ouzel, 42.
Natter-jack, the, 175.
Naturalist's World and the swallow, 76-7.
Nature and salmon-leaps, 255-6.
Naumann, and the reed-warbler, 28-9.

INDEX. 399

Neuroptera, the, 309–23.
Newman and the cocktail beetle, 305; and the Neuroptera, 309; and the stone-flies, 319–21; and the saw-fly, 349.
Newton, Professor, and the skylark, 82–3.
Newts, the, 176–7.
Nightingale, the, 96–9.
Nightjar, the, 161–5.
Nodding bur-marigold, the, 368.
Norwich and the sweet-flag, 369–70.
Notes and Queries, and the lapwing, 58.

O'BRIEN, Mr., and the swallow, 76–7.
Orange tag, the, 307.
"Ornithology of Shakespeare," *see* Harting.
Otter, the, 5–10.
Owl, the barn, &c., 157–61.
Ox-eye or great titmouse, 133.

PADDOCK toad, the, 174–5.
Pallas and the char, 276.
Parnell, Mr., and the salmon, 236, 253.
Paxton, Mr., and the salmon, 237.
Peacock-fly, the, 306.
Peewit or the lapwing, 56–8.
Pennant, and the wheat-ear, 112; and the frog, 172–3; and the toad, 174; and the perch, 195; and the roach, 222.
Pennell, Mr., and the rudd, 227; and the common trout, 237–8; and the salmon, 254–5; and the stone-fly, 319.
Perch, the, 193–7.
Peregrine Falcon, 156.
Perla marginata, the, 320.
Phillips, Mr., and the daisy, 360–1.
Phryganidæ, the, 316–18.
Pictet and the Ephemeridæ, 324, 331, 340–4.
Pied wagtail, the, 66–9.
Pike, Mr. A., and the salmon, 243–4.
Pike, the, 280–5.
Pindar, Peter, and the robin, 106–7.
Pipistrelle or flitter-mouse bat, 18–19.
Pipit lark, 84.
Plantain, the great water, 391.
Plott's "History of Staffordshire," 15.
Plume-moths, the, 354.
Pæderus riparius, the, 307.
Pollan, the, and the smelt, 278–9.
Pond-weeds, the 373–94.
Pope, the, or ruff, 197–8.
Postlethwaite and the nightjar, 164–5.
Primrose, the pale, 361.
Prussian carp, the, 218–19.

QUADRUPEDS, the (*see* Otters, Bats, &c., &c.), 1–20.
Quarles, F., and the cuckoo, 89.
Quekett, Mr. J., and the owl, 159.

RALSTON and the sparrow, 121.
Ray's "Wisdom of God in the Creation," Introduction, v.–vi. ; and the rook, 136–7, 144–5; and Salmonidæ, 232.
Redbreast, the, 105–8.
Redbreasted merganser, the, 53–4.
Red field-vole, the, 5.
Red mullet, the, 186.
Redpoll, the, 120.
Redstart, the, 113–14.
Red tag, the, 306.
Reed, the common, 388–9.
Reed canary grass, 389.
—— meadow-grass, 389.
Reed-mace, the, 389–90.
Reed-sparrow, the, 29.
Reed-warbler, the, 27–8; nest of, 69.
Rennie and the cuckoo, 89.
Reptiles, the (*see* Lizards, &c.), 166–77.
Ring-dove, the, 149.
Ringed snake, the, 170.
River bull-head, the, 200–1.
River-trout, the, 262–70.
Roach, the, 222–3.
Robertson, Mr., and salmon, 252–4.
Robin redbreast, the, 105–8.
Rogers and the redbreast, 108.
Ronalds, and insects, 306–23; and the Ephemeridæ, 344–6; and saw-fly, 349; and flies, 352.
Rook, the, 136–46.
Rooper, Mr. G., and the water-ouzel, 43–4; and young salmon, 240.
Rope, Mr. G. T., and water-rat, 3–4; and the weasel, 11–12.
Rudd, the, 224–5.
Ruff, the, or pope, 197–8.
Russel, Mr., and the salmon, 237.

SAILOR-FLY, the, 307.
St. James' Gazette and the night-jar, 163.
St. John, Mr., and the fly-catcher, 112–13.
Salmon, the, 234–58; and Mr. G. M. Kelson, 183–4, 186.
Salmonidæ, the, 231–34.
Salter and the barbel, 211; and the smelt, 277.
Sand-fly, the, 319.
Sand-lizard, the, 166–8.
Sand-martin, the, 79.
Sandpiper, the common, 65–6.
Satin moth, the, 353.
Saunders, Mr. Howard, *see* Yarrell.
Saw-flies, the, 348–9.
Schrœder and the swallows, 74–5.
Scorpion-flies, the, 319.
Scorpion, the water, 350.
Scott, Sir W., and the heron, 63; and the curlew, 152–3; and the owl, 161.
Scottish frog, the, 171.
Screech owl, the, 158.

INDEX.

Scrope, Mr., and salmon and grilse, 250-1, 255.
Sea-lamprey, the, 292.
Scaly, Mr., and eels, 289-90.
Sea-trout, the, 258-62.
Sea-trout parr, the, 238-9.
Sedge, the great, 365.
Sedge, the red, &c., 318.
Sedge-warbler, the, 25-7.
Seebohm ("British Birds"), and the sedge-warbler, 27; and the reed-warbler, 27-8; and bearded titmouse, 30; and moor-hens, 32; and the water-rail, 37; and the water-ouzel, 39; and the mallard, 47; and the snipe, 60; and the wagtail, 69; and the cuckoo, 91; and the white-throat, 102; and the redstart, 114; and the whinchat, 116; and the titmouse, 129-33; and the sparrow-hawk, 154; and the kestrel 155; and the owl, 158.
Shad, the, 229-30.
Shad-bird, the, see Sandpiper.
Shakespeare, Wm., and the kingfisher, 25; and the lapwing, 57-8; and the martin, 78; and the skylark, 80; and the nightingale, 98-9; and the redbreast, 106; and the hedge-sparrow, 115; and the kestrel, 156; and the owl, 158-60; and the shardborne beetle, 305; and the cowslip, 362; and daffodils, 363.
Shaw, Mr., and salmon, 238-47.
Shepster or starling, 85-6.
Shrew-mouse, the, 14-17.
Shrimp, the, 298.
Silver-horns, the, 318-19.
Siskin or aberdevine, 120.
Skelton, and the coot, 34; and the sparrow, 121.
Skylark, the, 80-4.
Slow-worm, the, 168-9.
Slugs, 293.
Smelt, the, 277-9.
Smith, Mr. G., and the salmon, 249.
Snails, 293-6.
Snakes, 169-71.
Snipe, the, 59-60; the summer, 65-6.
Song-thrush, the, 92-4.
Sowerby's "English Botany," 359.
Spalding, Mr., and Loch Leven trout, 269.
Spallanzani and bats, 20.
Sparrow, the house, 121.
Sparrow-hawk, the, 153-4.
Speedy, Mr., and the rook, 138-41.
Spence, and the dragon-fly, 314, 317-18; and ants, 347-8.
Spenser, and the owl, 158.
Spider-fly, the, 352.
Spiders and gossamer-web, 354-5.
Spined loach, the, 213.

Spotted fly-catcher, the, 112-13.
Stockhouse, Mr., and the pond-weeds, 373.
Starling, the, 85-6.
Stephens and the cock-tail beetle, 305; and the Ephemeridæ, 324, 327, 331, 340-6.
Sticklebacks, the, 201-3.
Stoat, the, or ermine weasel, 12-14.
Stoddard and the salmon, 255.
Stoddart and worms, 357-8.
Stone-fly, the, 319-21.
Swainson, and the skylark, 81-2; and the cuckoo, 89; and the blackbird, 96; and the white-throat, 102; and the chaffinch, 120; and the sparrow, 121; and the night-jar, 164.
Swallow, the, 73-7.
Swan, the, 49-53.
Sweet-flag or sweet-rush, 368-70.
Sweet-violet, the, 363.
Swift, the, 70-3.

TAVERNER ("Experiments on Fish,' &c.) and the frog, 173.
Tawny owl, the, 160-1.
Taylor, John, and the wheat-ear, 112; and the salmon, 254.
Teal, the, 54-5.
Tegetmeier, Mr., and the titmouse, 132.
Tench, the, 215-18.
Tennyson and the swan, 52.
Thompson ("Natural History of Ireland") and the salmon, 236-7.
Thomson and the swan, 50; and the swallow, 75.
Thrumwort, the great, 391.
Tiger-moth, the, 353.
Tissue-moth, the common, 353.
Titlark or meadow-pipit, 84.
Titmouse, the, 129-35.
Toads, the, 174-5.
Treat, Mrs., and the bladderwort, 370.
Tree-pipit or pipit-lark, 84.
Treherne, Major, and the sea-trout, 259.
Trichoptera, the, 315-16.
Troughton, Mr., and the otter, 8.
Trout-parr, the, 238-9.
Tuberville's sonnet, 49.
Turner, Dr. W., and the water-hen, 33.
Twaite-shad, the, 229-30.

VALERIAN, the great wild, 368.
Venus's catch-fly, the, 363.
Violet, the, 362-3.
—— the water, 370.

WAGTAIL, the water, 66-9.
Walton, I., and the water-rat, 2; and the otter, 6; and the nightingale, 96; and the frog, 172-3; and the carp, 207; and the barbel, 210; and the gudgeon, 211; and the tench, 215; and the bream, 220; and the chub, 224; and the rudd,

INDEX.

224; and the pike, 281; and the black slug, 293.
Water-agrimony, the, 392.
Water-beetle, the great, 302.
Water-betony, the, 383.
Water-boatman, or boat-fly, 302-3.
Water-bug, the, 303.
Water-cress, the, 382-3; the fool's, 390.
Water-cricket, the, 349-50.
Water-crowfoot, the, 367.
Water-cup, the, 365-6.
Water-dock, the great, 386.
Water-dropwort, the, 384.
Water-ermine moth, 353.
Water-figwort, the, 383.
Water-gladiole, the, 375.
Water-hemlock, the, 390.
Water-hen, the, 31-3.
Water-horehound, the, 378-9.
Water-horsetail, the great, 375-6.
Waterhouse, Mr., and the water-vole, 1.
Water-lily, the common, 384-5.
—— the lesser, 389-1; the fringed, 384; the large white, 385-6.
Water-ouzel, the, 38-44.
Water-parsnep, the broad-leaved, &c., 390-1.
Water-plantain, the great, 391; the floating, 391.
Water-rail, the, 36-8.
Water-scorpion, 350.
Water-shrew, the, 16-17.
Water sweet-grass, 367.
Water-thyme, the, 393-4.
Water-violet, the, 370.
Water-vole or water-rat, 1-5.
Water-wagtail, the, 66-9.
Weasel, the, 10-12.
Weaver, Rev. T., and the carp, 207; and the tench, 215; and the bream, 221.
Westwood, and the cock-tail beetle, 305; and the Neuroptera, 309; and the Phryganidæ, 316-17; and the Diptera, 351.
Wheat-ear, the, 110-12.
Whinchat, the, 115-16.
Whirligigs or *Gyrinus natator*, 301-2.
Whitaker, M. J., and the wren, 104-5.
White, Gilbert, and the weasel, 12; and the shrew-mouse, 15; and the swift, 72; and the swallow, 75; and the black-cap, 109-10; and the wheat-ear, 112; and the rook, 144; and the nightjar, 161; and the carp, 207; and the water-bug, 303; and the dor-beetle, 306; and the gossamer-web, 354-5.
White-rot, the, 365-6.

White-throat, the, 101-2.
Whorl-grass, the, 367.
Wild-duck or the mallard, 46-9.
Williamson, Capt. T., and salmon-leaps, 256; and the smelt, 277-8.
Willoughby, and the swallows, 74-5; and the dotterel, 150-1; and the kestrel, 156.
Willow-flies, the, 321-2.
Willow-herbs, the, 387.
Willow-warbler, the, 99-101.
Willow-wren, the, 99-101.
Wilson, Mr., and the bladderwort, 370.
Windhover, the, 155-6.
Woodpeckers, the, 124-8.
Woodpie, the, 127-8.
Wood-pigeon, the, 149.
Wood-sorrell, the, 363.
Wordsworth, and the wren, 104; and the redbreast, 107; and the hedge-sparrow, 115; and the blue tit-mouse, 135; and the dor-hawk, 162.
Worms, the earth, 356-8.
Wotton and the swallow, 75; and the martin, 78.
Wren, the, 102-5.
Wren-tail, the, or frog-hopper, 350.
Wry-neck, the, 123-4.

YARRELL ("British Birds," 4th edition), and kingfisher, 23; and the swan, 50-2; and the heron, 62; and the sandpiper, 66; and the swift, 70; and the cuckoo, 86-90; and the wry-neck, 123-4; and the coal-tit-mouse, 130-1; and the perch, 193, 195-7; and the pope, 197-8; and the river bull-head, 200; and the sticklebacks, 201-3; and the carp, 207, 219; and the gudgeon, 212; and the loach, 213-14; and the rudd, 225; and the dace, 226; and the salmon, 235, 255; and the river-trout, 263-9; and the smelt, 277; and eels, 286.
Yellow-hammer, the, 122-3.
Yellow sally, the, 321.
Yellow water-iris, the, 376-7; and the frog-bit, 381.
Young, Mr., and salmon, 241-2; and grilse, 250.

"ZOOLOGIST," the, and water-vole, 3-4; and the weasel, 11-12; and the water-ouzel, 43; and the dabchick, 45; and wild-duck, 49; and the wren, 104-5; and the nightjar, 164-5; and the tench, 216.

PRINTED BY BALLANTYNE, HANSON AND CO.
EDINBURGH AND LONDON.

www.ingramcontent.com/pod-product-compliance
Lightning Source LLC
Chambersburg PA
CBHW030557300426
44111CB00009B/1018